YOUR BABY'S
MICROBIOME

T .... aginal Birth *and*

Breastfeeding *for* Lifelong Health

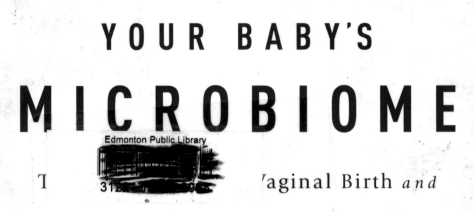

TONI HARMAN
and ALEX WAKEFORD

From the Directors of the Award-Winning

Documentary MICROBIRTH

Praise for *Your Baby's Microbiome*

"Toni Harman and Alex Wakeford shine the brightest of spotlights on the importance of birth and early infancy. *Your Baby's Microbiome* is compelling and informative—a must read for parents-to-be."
—**Dr. Rodney Dietert**, Professor of Immunotoxicology, Cornell University; author of *The Human Superorganism*

"You thought you were human. But you're actually a multi-species ecosystem. We all cohabitate with bacteria, fungi, protists, and even microscopic animals, and scientists are discovering that the way we treat the microbial companions that live in us and on us has a lasting impact on our health. Exploring what we know about the role symbionts play in childbirth and the early days of a baby's life, *Your Baby's Microbiome* is a fascinating read for anyone interested in what it means to be human." —**Jennifer Margulis**, PhD, award-winning journalist; author of *Your Baby, Your Way*; coauthor, with Paul Thomas, MD, of the bestselling *The Vaccine-Friendly Plan*

"For expectant families, this is the must-read book of this generation. As an advocate for parents, focusing on how the maternal environment changes the long-term health outcomes for the baby, this book is adding a critical piece to the puzzle of health for our children. The 'seeding and feeding' of a child's microbiome could potentially be one of the most important lessons for parents. This timely information should be included as part of every childbirth education and newborn care class. *Your Baby's Microbiome* will help parents truly make informed decisions about how and where they give birth and how they feed and care for their baby. It is a must for every woman giving birth." —**Laurel Wilson**, IBCLC, CLE, CLD, CCCE, coauthor of *The Attachment Pregnancy*

"Our expanding understanding of the roles of the microbiome and epigenetics in health and disease is profoundly changing maternity care and public health for the better. The conversational style and clear explanations in *Your Baby's Microbiome* by some of the world's leading scientists and maternity care providers make this new and complex information accessible and inspiring to the general public. We now know ways to support the role of healthy microbes in our bodies and in our environment to improve our lifelong health and well-being." —**Penny Simkin**, author, doula, and birth educator

"How we give birth and feed our babies should be choices, not something for which we have to fight for support. A huge 'thank you' to Toni Harman and Alex Wakeford for adding to a critical perspective on the individual and global impacts of some of those choices, in hopes that this conversation will keep going—pushing our maternity care systems to support physiological processes, as well as medicalized ones." —**Cristen Pascucci**, founder, Birth Monopoly; cocreator, Exposing the Silence Project; vice president, Improving Birth (2012–2016)

"*Your Baby's Microbiome* is essential reading for every expectant parent, grandparent, and anyone who works with or cares about childbirth, and the health and well-being of the next generations. It opens the door to what our intuition already knows—that disturbing nature's well-orchestrated design in childbirth has short- and long-range health risks. You can make a difference and turn this potential health disaster around. Knowledge is power, and Toni and Alex have put the power into your hands." —**Debra Pascali-Bonaro**, founder and president, Pain to Power; director of the award-winning documentary *Orgasmic Birth: The Best-Kept Secret*; coauthor of *Orgasmic Birth: Your Guide to a Safe, Satisfying, and Pleasurable Birth Experience*

"I hope this extremely useful and informative book will be widely distributed and widely read, and that its findings will change birthing practice in this country and around the world!" — **Robbie Davis-Floyd**, PhD, senior research fellow, Department of Anthropology, University of Texas Austin; author of *Birth as an American Rite of Passage*; lead editor of *Birth Models That Work*

YOUR BABY'S
MICROBIOME

YOUR BABY'S
MICROBIOME

The Critical Role *of* Vaginal Birth *and*

Breastfeeding *for* Lifelong Health

TONI HARMAN
and ALEX WAKEFORD

Chelsea Green Publishing
White River Junction, Vermont

Originally published in the United Kingdom by Pinter & Martin Ltd in 2016 as *The Microbiome Effect*. This edition published by Chelsea Green Publishing, 2017.

Editor: Judy Barratt
Project Editor: Brianne Goodspeed
Copy Editor: Angela Boyle
Proofreader: Laura Jorstad
Indexer: Ruth Satterlee
Designer: Melissa Jacobson
Page Composition: Abrah Griggs

Printed in the United States of America.
First printing February, 2017.
10 9 8 7 6 5 4 3 2 1   17 18 19 20 21

Our Commitment to Green Publishing
Chelsea Green sees publishing as a tool for cultural change and ecological stewardship. We strive to align our book manufacturing practices with our editorial mission and to reduce the impact of our business enterprise in the environment. We print our books and catalogs on chlorine-free recycled paper, using vegetable-based inks whenever possible. This book may cost slightly more because it was printed on paper that contains recycled fiber, and we hope you'll agree that it's worth it. Chelsea Green is a member of the Green Press Initiative (www.greenpressinitiative.org), a nonprofit coalition of publishers, manufacturers, and authors working to protect the world's endangered forests and conserve natural resources. *Your Baby's Microbiome* was printed on paper supplied by Thomson-Shore that contains 100% postconsumer recycled fiber.

Library of Congress Cataloging-in-Publication Data
Names: Harman, Toni, author. | Wakeford, Alex, author.
Title: Your baby's microbiome : the critical role of vaginal birth and breastfeeding for lifelong health / Toni Harman and Alex Wakeford.
Description: White River Junction, Vermont : Chelsea Green Publishing, 2017. | Includes
  bibliographical references and index.
Identifiers: LCCN 2016042186| ISBN 9781603586955 (paperback) |
  ISBN 9781603586962 (ebook)
Subjects: LCSH: Human body--Microbiology. | Bacteria--Health aspects. | Pathogenic microorganisms. | Immune system--Children. | BISAC: HEALTH & FITNESS / Pregnancy & Childbirth. | SCIENCE / Life Sciences / Biology / Microbiology. | HEALTH & FITNESS / Diseases / Immune System. | MEDICAL / Microbiology. | SCIENCE / Life Sciences / Human Anatomy & Physiology. | SCIENCE / Life Sciences / Biology / General. | HEALTH & FITNESS / Diseases / Abdominal.
Classification: LCC QR171.A1 H37 2017 | DDC 579--dc23
LC record available at https://lccn.loc.gov/2016042186

Chelsea Green Publishing
85 North Main Street, Suite 120
White River Junction, VT 05001
(802) 295-6300
www.chelseagreen.com

To our daughter, Willow Lula

Contents

Introduction

This book is about the microscopic events that happen during childbirth. The latest scientific research reveals that these events could have ramifications for a child's lifelong health, and they may even impact the future of our species.

To explain this grandiose opening statement, first we need to introduce one of the hottest topics in science right now: the human microbiome.

If you want it to be a surprise as we reveal the latest scientific discoveries chapter by chapter, feel free to skip forward a few pages. Alternatively, if time is short for whatever reason (like you're about to have a baby), just read the next three pages and cram in as much information as you can.

There have been thousands of scientific papers published in recent years about the community of microorganisms that live on and in your body. There have been several hundred media articles and quite a few factual TV programs that mention the incredible secrets of the microbiome. There have even been a couple of full-length documentaries, including our own film, *Microbirth*. This book is based on the 3 years of extensive research we did for our film, and included in this book are excerpts from transcribed interviews with twelve leading professors.

What is the microbiome?

As we'll discuss in the first chapter, the human microbiome comprises trillions of microorganisms that live on you and in you—the bacteria, fungi, viruses, protozoa, and archaea. These microbes live on your skin, in your gastrointestinal tract, in your urogenital system, in your mouth, in your nose, in your lungs, and, if you're a woman, inside your vagina.

The most well-studied population of microbes lives in your gut. (As the human microbiome is made up mainly of bacteria, we'll use the words *microbes* and *bacteria* interchangeably throughout the book.) These trillions of microbes play a very important role: They keep the body functioning and help protect you from disease.

Scientists have discovered that the single most-critical moment in terms of founding the human microbiome happens in the narrow time period around childbirth. There may be some prenatal exposure to microbes during late pregnancy, and there's a whole lot more happening with breastfeeding, but the main seeding event for the baby's gut microbiome occurs around birth. Amazingly, these microscopic events could determine the child's health for the rest of his or her life.

In chapter 2 of our book, Dr. Martin Blaser, director of the human microbiome program at New York University and author of *Missing Microbes*, will describe how the fetus develops in a near-sterile environment in the womb, which means that childbirth is likely to be the baby's first main contact with the world of microbes. During vaginal birth, as he or she travels through the birth canal, the baby is coated in the mother's bacteria. This payload of microbes goes into the baby's eyes and ears, up its nose, and into its mouth. Inevitably, the baby swallows some of them, too.

In the baby's gut, the first bacteria to arrive start to colonize and multiply. Special breast-milk sugars, called oligosaccharides, are indigestible to the baby, but they are there purely to feed the baby's newly seeded gut bacteria. This natural "seed-and-feed" process is perfectly designed to set up the baby's gut microbiome in the optimal way.

Seed and feed. The two amazing sides of the founding of the human microbiome.

The latest research indicates that this seed-and-feed process could be critical for the development of the infant immune system. Emerging science suggests that the first bacteria to arrive in the baby's gut plays a critical role in the training of the immune system, helping it to identify what is friend and what is foe (in other words, which bacteria the body should tolerate and which it should attack). Interfering with this process could result in the incorrect training of the baby's immune system, in turn resulting in the immune system attacking beneficial bacteria and tolerating harmful bacteria. Overall, this inadequate training potentially sets a pathway for health problems later in the child's life.

Just as the baby develops from a newborn to a toddler, so the baby's microbiome develops over the first few months and years of life until the microbiome stabilizes sometime during early childhood.

How does a Cesarean section affect a baby?

In chapter 4, we'll explore cutting-edge research that looks at if and how it might be possible to partially restore the microbiome of a baby born by Cesarean section (C-section) with vaginal microbes immediately upon its arrival in the world. Because a Cesarean section baby hasn't traveled all the way through the birth canal, the first microbes he or she come into contact with (aside from possible prenatal exposure) are likely to come from a number of sources within the environment of the operating theater, not from the mother's vaginal tract. Sources include the skin of other people present in the operating theater—the mother, father, surgeon, anesthesiologist, midwife, nurse, or anyone else who happens to be present at the time of the birth.

Dr. Rodney R. Dietert, author of *The Human Superorganism*, describes the microbiome of babies born by Cesarean section as being "incomplete." In his 2012 paper "The Completed Self," Dr. Dietert describes how babies born by Cesarean section may not receive the full complement of microbes that they are supposed to receive in the narrow window surrounding birth.[1]

In a later award-winning paper, "The Microbiome in Early Life," published in 2015, Dr. Dietert went as far as to say that those babies who did not have optimal seeding of their microbiome at or near birth have "the functional equivalent of a birth defect."[2]

3

Then there's the potential for epigenetic change associated with Cesarean section. As we'll learn in chapter 6, particular environmental exposures can switch genes on or off. Right now, scientists are investigating whether or not the mode of birth—whether by Cesarean section or vagina—is just such an environmental exposure. Could it be that, without the stresses, pressures, and hormones associated with vaginal birth, some genes relating to life outside the womb are somehow not switched on or off at all, or are not switched on or off at the right time?

If research proves that this is true, what ramifications are there for a Cesarean section baby throughout his or her life, and what effects might that one birth have generations later?

What is the impact to individual and global health?

Whether as a result of changes in the baby's gut microbiome, the impact of epigenetics changes childbirth, a combination of these two factors, or indeed another reason altogether, research shows that a baby born by Cesarean section is at significantly increased risk of developing certain health conditions later in life, including asthma, type 1 diabetes, celiac disease (a disease of the intestine related primarily to gluten intolerance), and becoming obese.

These are examples of noncommunicable diseases (NCDs), diseases that cannot be passed from one person to another. And according to Dr. Dietert, they could be just the tip of the iceberg. He says that if you have one NCD early in life, you are at increased risk of developing other diseases later in life, including cardiovascular disease, other autoimmune disorders, bowel problems, and even some cancers. NCDs are already the world's number one killer and, as we'll see in chapter 8, by 2030 the cost of treating them may bankrupt our global healthcare systems.

But there's something else. According to Dr. Blaser's hypothesis in *Missing Microbes*, the current plague of disease in industrialized nations could be related to the depletion of bacterial diversity in our gut microbiome. Antibiotics, modern diet and lifestyle, and an increase in the numbers of Cesarean section births could all contribute to this reduced diversity. As we'll see in chapter 8, Dr. Blaser's worst-case scenario is an "antibiotic winter" in which we've all become more susceptible not only to NCDs but also to infectious diseases. Globally, the way we live our lives today may be increasing the likelihood of pandemics.

While this particular vision of the future looks pretty bleak, the discovery of the importance of the microbiome and epigenetics has, we hope, come just in time to turn things around. Our new understanding gives us all the opportunity to take a different path. The future is not just in the hands of scientists. It's in our hands, too.

What can we do?

Promote the information in this book. Champion more research. Call for more training of healthcare professionals. And start living a life more in harmony with our microbes.

This book reveals the latest thinking on how to optimally seed and feed a baby's microbiome in order to build the strongest possible immune system. We think this is something that all expectant parents, doctors, midwives, doulas, birth educators, and even teenagers (who will become the parents of tomorrow) need to know about.

Who are we?

We are Alex and Toni, a couple of filmmakers, who met 20 years ago at the London Film School. We were drawn to each other through a dream to make feature films, in particular psychological thrillers. We joined forces and formed a production company, Alto Films Ltd. (the "Al" from Alex, the "to" from Toni).

In 2006, we made our first feature film, *Credo*, a spooky thriller that was picked up by Lionsgate and released worldwide as *The Devil's Curse* in 2008. We immediately followed this up with our next production, our baby daughter called Willow. It was her birth by emergency C-section that changed our view on life and ignited a new and unexpected passion for the miracle of childbirth.

Our next three films explored different perspectives of childbirth in the twenty-first century: the role of doulas in pregnancy, birth, and postnatal care in *Doula!* (2010, 60 mins.);[3] the politics of human rights in childbirth in *Freedom for Birth* (2012, 60 mins.);[4] and most recently the science of childbirth in *Microbirth* (2014, 60 mins.).[5]

Microbirth is a science documentary looking at birth in a whole new way, through the lens of a microscope. The film won the top prize—the Grand Prix Award—at the Life Sciences Film Festival in Prague in October 2014, and it is now being distributed internationally.

How did we assemble the team?

Think of a classic heist movie. Someone wants to get their hands on a priceless diamond, but they can't pull it off on their own. They need help. They need to assemble their A-team of experts. Same for us.

In our heist movie, as filmmakers/authors, we are the lead characters. We have a goal: to discover the long-term implications of how we give birth today. That's our priceless diamond.

The inspiration for this whole adventure happened by chance in June 2011. In movie parlance, here's our backstory. And in movie-script format, here's how it all began . . .

FADE IN:

EXT. HILLSIDE HOTEL, NORTH OF ENGLAND—DAY

Evening sunlight falls on a gothic hotel nestled in the hillside overlooking a small coastal town.

CUT TO:

INT. HILLSIDE HOTEL, MEETING ROOM—DAY

INTRODUCTION

CAPTION: NORMAL LABOUR AND BIRTH CONFERENCE 2011, GRANGE-OVER-SANDS, ENGLAND

Under the glare of movie lights, two people sit opposite each other in a small side room, an interviewer, TONI HARMAN, and the interviewee, SOO DOWNE, professor in midwifery studies at the University of Central Lancashire.

ALEX WAKEFORD hits record on a small video camera and nods to TONI HARMAN to start the interview.

> TONI HARMAN
> Tell me about your current research project.

> SOO DOWNE
> I'm starting to look at how mode of birth could be linked to negative health outcomes later in life. We think one of the mechanisms for this could be epigenetics.

TONI HARMAN scratches her head and tries to think of something clever to say.

> TONI HARMAN
> Oh . . . Um . . . Er . . . Epi-what?

> SOO DOWNE
> Epigenetics is the study of the expression of a gene. It could be that how a woman gives birth switches on or switches off certain genes in her baby. If a woman doesn't go through normal physiological birth, then there could be epigenetic changes happening in the baby which could affect her child's health later in life.

TONI HARMAN sits up straight. Her eyes light up.

                                        CUT TO:

INT. CAR—NIGHT

Rain lashes the windshield. ALEX WAKEFORD is
driving. TONI HARMAN is in the passenger seat.

                    TONI HARMAN
    Out of the dozen or so interviews we shot
    today, which one stood out the most?

                    ALEX WAKEFORD
    Professor Downe and epigenetics.

                    TONI HARMAN
    Did you understand a word of it?

                    ALEX WAKEFORD
    No—(PAUSE)—but I think we've just found our
    diamond.

FADE OUT

But for now, the diamond would have to wait as we were in the
midst of making a film about human rights abuses in childbirth.
Fast-forward to the autumn of 2012.

We had just released *Freedom for Birth*, and we were itching to
return to our diamond. We wanted to make a film that would ignite
our own passion, that was important on a global scale, and that we
could make on our own on a modest budget.

We remembered that original interview with Professor Soo
Downe, and we started thinking about making a film about the

incredible emerging story of epigenetics. Professor Downe connected us with other academics in the newly formed Epigenetic Impact in Childbirth (EPIIC) international research group. We were excited. We thought we were onto something.

We set up interviews with US professors Sue Carter and Aleeca Bell, who were both in the United Kingdom attending a conference on vasopressin (one of the hormones of birth alongside oxytocin). We asked them, "What is the most exciting thing about epigenetics in childbirth today?" Their eyes sparkled as they announced, "The most exciting thing about epigenetics in childbirth is . . .

"We just don't know."

Now we realize that, for a scientific researcher, "we just don't know" is exciting. It means that there are new answers to discover. More questions to ask. More avenues for research funding.

For us as filmmakers, "we just don't know" means there's no film. Or at least, that that particular area of research is so new there isn't a film to be made about that topic—yet. For epigenetics and birth, it was too early. We needed to find a new diamond and a new question. If you hit on the right question, you can open up more questions and find yourself on an epic quest for answers.

Meanwhile, one of the EPIIC cofounders, Professor Hannah Dahlen from Western Sydney University, Australia, suggested we look into the latest research linking the microbiome with birth. That was the first time we had heard of the word *microbiome*. We typed it into Google and searched for videos.

Up came a link to a YouTube video featuring someone called Dr. Blaser. That video blew our minds.[6] Our question then became, "What has the microbiome got to do with birth?"

We came across a paper by Dr. Dietert. In his paper titled "The Completed Self," Dr. Dietert linked mode of birth with an increased risk of children developing certain health conditions later in life. That was a lightbulb moment for us.

A key decision we made early on was to go straight to the scientists. We wanted to hear directly from them about the findings of their research. We wanted the science to be beyond scrutiny. We didn't want anyone to be able to question the background, experience, integrity, and academic record of any of the academics.

We sent Dr. Dietert an email asking if we could interview him. He shot back a response straightaway. He said yes. That was the moment we committed wholeheartedly to this project, investing our own money and dedicating the next few years of our lives to it.

Several months later, when Dr. Dietert drew his Cesarean section "Tree of Disease" on a blackboard in an empty lecture theater (see page 131), we had goose bumps. We realized what we were doing was much more important than we had first thought. We weren't just talking about birth. We were talking about a whole trajectory of human health.

Two more key members of the assembled A-team of experts were microbiologists: the aforementioned Dr. Blaser, and Dr. Maria Gloria Dominguez-Bello, both from New York University. Their labs are side by side on an upper floor of a military veterans' hospital in Manhattan. On entering the building, we were frisked for weapons and grilled with security questions until, at last, we were given permission to enter. As we were escorted through a maze of corridors and alarmed doors, we cast uneasy glances at the numerous biohazard warning signs. It felt like we were entering secret facilities, access to which would ordinarily be denied to members of the public. Going back to our film school days when we wanted to make thrillers together, it felt as though we were actually on the set of a real-life sci-fi thriller. We felt very privileged to be in such a situation, and it reaffirmed that what was about to be revealed to us was likely to have a global impact.

Around the same time, we came across an article about Canadian research on the bacterial footprint of infants born by Cesarean section.[7] The article talked about a pilot study that found that babies born by Cesarean section may have different microbial profiles than babies born vaginally. We contacted one of the authors of the paper, Professor Anita Kozyrskyj, and several months later we flew to the foothills of the Canadian Rockies to interview her in Edmonton, Alberta.

We needed other experts in genetics and epigenetics, and we found them in Dr. Jacquelyn Taylor, assistant professor from Yale University, and Professors Carter and Bell, whom we had already interviewed.

We were starting to construct a network of experts from around the world, traveling thousands of miles to do so. We were drawing together their expertise and knowledge to form a picture of childbirth that was very different from the conventional one with which we had started out.

To provide a medical perspective, we found Dr. Neena Modi, a professor of neonatal medicine, after she had published a paper (with her research colleague Dr. Matthew Hyde) on the link between Cesarean sections and increased risk of children becoming obese later in life.[8] The next member of the team was Emeritus Professor Philip Steer, an obstetrician and former editor of the *British Journal of Obstetrics and Gynaecology* (now *BJOG: International Journal of Obstetrics and Gynaecology*).

We also needed the perspective of midwives. As well as Professor Dahlen, we interviewed Lesley Page, visiting professor of midwifery at King's College London and also current president of the Royal College of Midwives.

The final expert we needed was someone to offer an economic health perspective. Picking our daughter up from a school friend's birthday party, we fell into conversation with another parent who exclaimed, "I have exactly the right person for you."

"Who?" we asked with growing excitement.

The other parent smiled. "Turn around. He's standing right behind you." And that's how we met Stefan Elbe, professor of international relations at Sussex University and director of the Centre for Global Health Policy. That chance encounter completed the team.

That was it. We had found a story with global importance, and we had found our experts:

MARTIN BLASER, director of the human microbiome program and professor of translational medicine New York University; author of *Missing Microbes*—world-leading microbiome expert

MARIA GLORIA DOMINGUEZ-BELLO, associate professor in the Department of Medicine, New York University—expert in infant microbiome research

RODNEY DIETERT, professor of immunotoxicology, Cornell University; author of *The Human Superorganism*—provider of brilliant big-picture analysis on global health research and expert in the human immune system

ANITA KOZYRSKYJ, professor in the Department of Pediatrics, University of Alberta; coprincipal investigator at the Synergy in Microbiota Research (SyMBIOTA)—expert in infant microbiome research and immune development

SUE CARTER, professor; behavioral neurobiologist; director of The Kinsey Institute; Rudy Professor of Biology at Indiana University —world-leading expert in synthetic oxytocin

JACQUELYN TAYLOR, associate professor of nursing and assistant dean of diversity and inclusion, Yale University—expert in genetics and epigenetics

HANNAH DAHLEN, professor of midwifery, Western Sydney University —expert in epigenetic impact in childbirth and midwifery

ALEECA BELL, assistant professor in the Department of Women, Children, and Family Health Science, University of Illinois at Chicago—expert in epigenetic research in childbirth

NEENA MODI, professor of neonatal medicine, Imperial College London—medical and scientific expert on newborn babies

PHILIP STEER, emeritus professor of obstetrics, Imperial College London—medical and obstetric expert on Cesarean sections

LESLEY PAGE, visiting professor of midwifery, King's College London; president of the Royal College of Midwives—expert on midwifery and how physiological birth is supposed to work

STEFAN ELBE, professor of international relations, University of Sussex; director of the Centre for Global Health Policy—expert in how this cutting-edge science translates in the real world

One of the things that frustrated us in making our *Microbirth* film was that, by the very nature of the hour-long documentary format, you can only paint in broad brushstrokes. You lose the fine detail. Hour-long interviews become reduced to thirty-second sound bites. Complex science is simplified to make it as accessible as possible, and there's limited opportunity for nuance.

When we made *Microbirth*, more than 95 percent of what we shot was thrown away (well, technically, the rushes are all saved in dozens of digital hard drives). Writing this book, though, gives us a brilliant opportunity to unleash that missing 95 percent, to delve deeper, to give even more information, and so to paint a fuller picture of the amazing pioneering science that has the potential to make a real difference to people's lives.

Our own personal story has lasted several years and seen us travel from Europe to North America and back, twice over. Now we feel like we've done what we set out to do. We achieved our goal, we pulled off our "heist," and the information within this book represents our "priceless diamond."

Of course, we couldn't have pulled it off without the cooperation, help, and trust of our treasured A-team of expert professors. We are very grateful to all of them for trusting us to share their research findings. For us it is an absolute honor and a real privilege to share their wisdom with the world.

How should you use this book?

You don't need a science background to read this book. We wanted to make the science easy to understand and accessible to all but without undermining it in any way. We used ourselves as a benchmark. If we could understand the science, we felt that other laypeople would be able to understand it, too. We've worked with all the professors to make sure that we're offering you something well researched and properly explained.

We don't want to clutter up the text with unwieldy descriptions and titles for our experts, so we've simplified them whenever we can. Dr. Blaser's full title, role, and institution is Dr. Martin J. Blaser, MD, Muriel G. and George W. Singer Professor Translational

Medicine, Director of the NYU Human Microbiome Program, former Chair of the Department of Medicine, and Professor of Microbiology at New York University School of Medicine. But we think using Dr. Martin Blaser, or Dr. Blaser, will help to make the science and arguments far more digestible.

We love a good visual analogy (you've already experienced the one that refers to our heist movie). In our everyday conversations, we constantly come up with analogies. For us, a visual analogy is a fun shortcut to express a complex idea.

Each chapter contains a QR code linking to a short video that has been edited specifically for our readers. QR code reader apps are available to download for free from Google Play for Android or iTunes for Apple devices. You don't need to use the QR codes to access the videos, though—we've put all the URLs at the end of the book, too (see page 162), and you just need to input these into your browser. These videos provide fun supplementary material to the written text. You don't have to watch them, but if you do, we hope you'll enjoy seeing the professors—and sometimes us, too! We use quotations from the interview transcripts in this book, but we want to give you the opportunity to see the experts speaking in person to help bring their words to life.

We very much appreciate and indeed celebrate that it's not just women who are able to give birth. Transgender individuals who were born as a female with a womb and female sex organs, but who have undergone, or are undergoing, a gender transition to a male, may also be able to give birth.

So that we don't have to say "women or transgender individuals" each time, and also "mothers or transgender birth-giving individuals," we simply refer to "women" and "mothers." We sincerely hope this doesn't offend anyone.

A final note on language. We don't want our use of language to get in the way of communicating an idea. Sometimes, we might use different expressions to describe the same process or event. For example, the seeding of the baby's microbiome is sometimes referred to as the microbial transfer from mother to baby in the birth canal.

We'll also refer to the transfer of the "special cocktail" of bacteria, or a microbial "payload," or the more technical term "inoculum."

As much as possible, we will try to avoid using technical scientific terms. However, there are a few that we feel we can't avoid.

Which word is best?

Like any field of research, microbiology has its own vocabulary to describe the bacterial world. Some words are interchangeable; some words are very specific.

As we are not microbiologists, here's a very approximate layperson's guide to some of the words that come up time and again in our interviews with our experts. What follows is our understanding of the key words associated with this field of research.

Microbiology: The study of microscopic organisms.

Microbe: All-encompassing term for a microscopic organism, otherwise known as a microorganism. It covers bacteria, viruses, protozoa, archaea, fungi, and so on.

Microorganism: Microscopic organism. Used interchangeably with *microbe* (see above). Sometimes used interchangeably with *bacteria*.

Bacteria: The most common microscopic organisms. It's worth remembering that all bacteria are microbes, but not all microbes are bacteria. Bacteria are microscopic and come in all different shapes. They are thought to be the first life-form to have ever existed on Earth, and they can be found in every habitat on the planet (and even on manned spacecraft). *Bacteria* is the plural form of a single bacterium.

Germ: A slightly old-fashioned way to describe a microorganism. It has a negative connotation, associating microorganisms/microbes/bacteria with causing disease. It's true that a few species of "germs" (microorganisms) can cause disease, but the vast majority are not harmful and, in fact, can be very beneficial to humans.

Virus: Another type of microbe. A virus is smaller than a bacterium and requires a living host to keep it alive.

Microbiome: A complex community of microbes and their genes.

Microbiota: Often used interchangeably with *microbiome* (see above).

Gut flora: A description of the complex community of microbes
that live in the gut. As science is discovering that it's not just
"flora" in the gut, this term is considered to be a little outdated.
Scientists more often refer to gut microbiota or gut microbiome.

Antigen: An "identity tag" on a cell. If your immune system
doesn't recognize the tag, it will launch an attack against that
particular antigen.

Pathogen: Anything that causes a disease, including a bacterium
(such as strep throat), a virus (such as hepatitis A, B, or C), or a
fungus (such as athlete's foot). A tiny percentage of bacteria are
pathogenic, meaning they can be harmful to humans, but the
vast majority are either harmless or beneficial to us as a species
(see *germ*, above).

Why did we write this book?

We believe in the importance of giving all expectant parents access
to the latest evidence-based research in order for them to make
informed choices about how they have their baby.

We also want to make clear that we absolutely do not intend to
question the decisions of any parent who is planning or has had a
C-section. Nor do we want to make anyone feel guilty about their
choices. Toni gave birth to our baby by C-section, so we know first-
hand that not all babies can or should (depending upon particular
medical circumstances) be born vaginally.

Our intention is to give future parents access to the latest scientific
research—research that could be critical for the long-term health of
their child. Whether a mother goes on to give birth vaginally or she
has a C-section, all parents should have access to the information
that will help them make fully informed decisions in their particular
situation. Our belief is that this information should also be available
to all healthcare providers engaged in supporting birthing women
so that they can understand and support the parents' decisions.

Right now, critical information that could be really useful to expectant parents and to healthcare providers is locked up in scientific papers accessible only to the world of academia. We want to take that critical information and release it to make it accessible to one and all.

Everything in this book is information we wish we had known 9 years ago when we had our daughter, so that we, too, could have made an informed choice. We can't change the past, but in writing this book we can hope to change the future: We can spread awareness of all the things we have learned on our own journey.

In the last chapter, we have written a strong "call to action." We want to do whatever we can to protect future generations. We hope that, when you've read our book, you'll hear our call and you'll want to join us. That way, we can all change the world together.

Toni Harman and Alex Wakeford

MEET THE AUTHORS

— ONE —

What Is the Human Microbiome?

We're going to tell a story that begins with birth. However, to understand how birth fits into the epic tale of life itself, we need to travel back in time to the dawn of life on Earth.

Four and a half billion years ago, planet Earth was born. For the next half a billion years, meteorites bombarded the new planet. Once this bombardment ended, the Earth's surface began slowly to cool and stabilize. A crust formed, creating a hot, rocky terrain, the planet's first solid rocks. As the Earth continued to cool, clouds formed, producing huge volumes of rainwater, which became the world's oceans. Landmasses started to form and, around the same time, along came life.

Despite highly toxic conditions, single-cell microorganisms appeared around three and a half billion years ago. These living microbes, such as bacteria, didn't just· appear and survive. They appeared and thrived. Ever since then, bacteria have successfully and completely colonized the planet. Species of bacteria have been found up the highest mountains, at the bottom of the deepest oceans, and even far up into the atmosphere.

Simple bacterial cells might have come first. Soon afterward (or possibly before), another type of single-cell organisms—archaea

—came on the scene. Bacteria and archaea are both prokaryotes—single-cell organisms that lack a true nucleus. Indeed, they might have evolved from a common ancestor. However, genetically and biochemically, they are quite different. To get a bit technical, bacteria and archaea have different ribosomal RNAs (rRNA). Archaea have three RNA polymerases, whereas bacteria only have one. There are also differences in their cell walls and membranes. (Virtually all life that we see around us today, including all plants and all animals, are eukaryotes, which have a nucleus and energy-creating mitochondria.)

Around 600 million years ago, there was a surge of activity with the development of simple animals, then fish, proto-amphibians (amphibian-like organisms), and land plants. By 400 million years ago, evolution was really motoring with the development of insects, seeds, and amphibians. Reptiles developed around 300 million years ago, and mammals arrived around 100 million years after that. Then came birds and flowers, with primates evolving a mere 60 million years ago. It's only in the past 2.5 million years that the *Homo* genus evolved. The mere whippersnappers that are modern humans have been around for only the last 200,000 years. Put in another way, humans have been around for only 0.004 percent of our planet's history.

What's the role of bacteria in evolution?

Throughout all these different evolutionary stages—from microbes all the way to the development of our own *Homo sapiens* species—bacteria have always been there. Over billions and millions and thousands and hundreds of years to the present day, bacteria have been part of all other life-forms.

In other words, since the dawn of time, all life-forms have dealt with the presence of bacteria by merging with these microorganisms to become new life-forms. The same goes for human beings. As we have evolved, we, too, have merged with bacteria—so much so that bacteria are very much present in our bodies and even in all our cells. It's the same with every other animal or plant on Earth. If you break open any cell, as well as cellular components inside, you'll find bacteria.

20

A theory called the endosymbiotic hypothesis, first proposed by evolutionary biologist Lynn Margulis, is described by Dr. Maria Gloria Dominguez-Bello, "Our own cells are indeed a composite of bacteria and other ancestral cells. We have some bacterial components in our own cells, which are the mitochondria. These are ancestral bacteria that fused with other cells."

In other words, the mitochondria that exist inside every cell in our bodies are descended from bacteria. Each mitochondrion could even be described as the "great-great-great-great-great-great-great-great-great-great grandchild" of a free-living bacterium. Once upon a time the free-living bacterium fused with, or was engulfed by, another cell. That cell benefited because the free-living bacterium became a mitochondrion, which produced much-needed energy. The mitochondrion benefited because it began living in a protected nutrient-rich home. What's more, bacteria aren't just confined to being inside our human cells. Trillions more bacteria live outside our human cells. Microbes fill, coat, and surround our bodies.

Put in another way, our bodies aren't made up of only human cells. They are also full of bacteria. In effect, the human body is a complex ecosystem made up of human cells and microorganisms existing together. So you could say we are all part human, part microbe. As Dr. Dominguez-Bello says, "When you understand that, you start viewing every single individual or species as a composite. So we are pretty much an ecosystem that can walk. . . ."

When you begin to think about it, you start to see yourself in a new way. On a philosophical level as products of evolution on a bacterial planet, we are not just human, we are all more than human. In fact, to be human is to be more than a single living organism. To be human is not to be a single "me" but a manifold "us."

So, what exactly is the human microbiome?

As director of the human microbiome program at NYU and author of the book *Missing Microbes*, Dr. Martin Blaser is one of the world's leading authorities on the human microbiome. He describes the human microbiome as "...all the organisms that live in the human body: bacteria, fungi, viruses, and so on. They're the organisms that live in and on us that make us home."

Up until recently it was estimated that there were one hundred trillion microbes compared with ten trillion human cells living in our bodies. This means our bodies were thought to be composed of 90 percent microbes and only 10 percent human cells. That's not in weight, but in number. In weight, all the microbes in a single body were estimated to weigh just over 1 kg (3 lbs.). Or to put it another way, about the weight of a human brain.

Scientists have recently revised this estimate—we know that we can't make generalizations about the ratio of microbes to cells because it depends upon a person's size and even perhaps what they have just eaten.[1] In fact, the ratio could be as great as 100:1 (100 times more microbes than human cells) or as low as 1:1 (the same number of microbes as human cells). Suffice to say that, whatever the ratio, there are a whole lot of microbes in your body!

However, the story doesn't end there: The thousands of different species of bacteria that live on and in our bodies carry a huge amount of genetic material.

Let's take a step back. There are estimated to be between 20,000 and 25,000 genes within the human body. Collectively, these genes are known as the human genome. And the human genome carries all the genetic instructions we need for human growth and development.

On top of that the microbes within our bodies also have their own genome. Researchers within the Human Microbiome Project "estimate that the human microbiome contributes some 8 million unique protein-coding genes or 360 times more bacterial genes than human genes."[2] If that estimate is correct, there could be several hundred times more genetic material within our microbes than there is carried within our human genes.

Lesley Page, elected president of the Royal College of Midwives at the time of writing, eloquently conveys the wonder of the knowledge that we are more microbe than human: "It's just as if we have discovered a whole new world, and it makes me change the way I think about myself. I used to be aware that I had organisms on my skin, and in my gut, and inside me, but to think I am more microbe than human cells, it's as if it connects me with the universe a bit more."

Where are the bacteria in your body?

Bacteria inhabit those parts of your body that come into contact with the outside world. They are found on the outside, on your skin. And they are found on the inside, in your mouth, throat, airways, and lungs, because you inhale quite a few every time you take a breath. Based on a constant rate of twelve breaths a minute, it's estimated that someone could inhale fifty bacteria with each breath. That's 600 a minute; 36,000 an hour; totaling more than 860,000 bacteria every day.[3]

There are bacteria in your eyes, ears, and nose—all of them "openings" to the outside world through which bacteria can enter your body.

The urinary tract (urethra and bladder) provides another opening through which bacteria can enter and colonize. Historically, urine was thought to be sterile (meaning there are no microbes present). However, recent advances in gene sequencing have shown that bacteria are present at low levels in the urine of healthy individuals.[4] Bacteria are not usually present in the kidneys of a healthy person, but if they develop cystitis or a urinary tract infection, bacteria can travel up the urethra, into the bladder, and upward to infect the kidneys. Research is ongoing as to whether or not bacteria are ordinarily present in the blood, heart, liver, pancreas, and ovaries of healthy individuals.

The brain was always thought to be sterile, but recent research has shown otherwise. Researchers looking at whether people with HIV/AIDS might be more prone to brain infections discovered that every brain they studied contained bacteria, regardless of the person's HIV status.[5] No one is quite sure how or when bacteria made it across the blood–brain barrier, but a microbial population is there.

Coming back to the location of microbes in our bodies: In women, there's a colony of microbes living inside the vagina. The vaginal microbiome is extremely important because it contains the reservoir of microbes for future generations. During pregnancy the placenta may also host small colonies of microbes. We'll explore the microbial colonies present in both the vagina and the placenta in the next chapter.

Critically, microbes are also found in every inch of your gastrointestinal (GI) tract (otherwise known as your gut)—all (roughly) 9 meters (30 feet) of it.[6] (The whole surface area of the digestive tract is thought to be the size of a football field.) Broadly, the GI tract covers everything from your mouth downward, including your stomach, small intestine, large intestine, and anus. Essentially, the GI tract is a tube with space in the middle where the food passes down. A thin, moist layer of cells called the mucous membrane, or the mucosa, lines the inside of the tube. The membrane consists of an epithelium (a layer of epithelial cells) attached to loose connective tissue (the lamina propria) underneath which is a layer of smooth muscle (muscularis mucosae) that helps food move along the gut. Together, these make up a gut barrier. This barrier separates the contents of the gut from the rest of the body.[7]

When people refer to "gut bacteria," they are usually referring to the many trillions of bacteria that live toward the end of the GI tract, in the large intestine. Many trillions of microbes live in, on, or around your gut lining. Interestingly, research has shown that whenever you have a bowel movement, gut bacteria could comprise up to 60 percent of the solid component of fecal matter.[8]

The gut microbiome is the most heavily studied microbial community of all. Partly this is because emerging science is indicating a strong gut microbiome–brain connection; what happens with the bacteria in your gut may affect brain development and behavior. However, as we'll see in chapter 5, the GI tract also plays a prominent role in the workings of the immune system.

24

Everybody's microbiome is different. According to Dr. Blaser, a person's microbiome "is as unique as their own fingerprint." Even identical twins have a unique microbiota profile. This is because from the moment we're born, we're constantly exposed to different species of bacteria. Every time you take a breath, every time you eat something, every time you smell something, and every time you touch something, you pick up microbes. You won't be able to see the billions of bacteria you're constantly ingesting and inhaling, but those bacteria may soon take up residence within your body, and many will soon become part of you.

What do the bacteria in our bodies actually do?

Like most living organisms, human beings have a symbiotic relationship with the microbes that live on and inside our bodies. In other words, the relationship between human cells and microbes is a two-way street. Your human cells do things that benefit the bacteria inside your body, and conversely the bacteria do things that benefit the human cells. This interaction—cells and microbes working together as a team—benefits the whole organism (you). You could say your human cells plus the bacteria, viruses, archaea, fungi, and protozoa are all part of the same team, namely "Team Human."

The adage that "teamwork works" really is true of your microbiome. The microbes inside you help keep your organs functioning and help protect you from disease. In return, you give the microbes a home and you provide them with food.

According to Dr. Rodney Dietert, together "we've formed what is called a symbiotic superorganism because we do things for each other and those things are very important for the health of the total organism."

Gut microbes help break down nutrients, so if those microbes are missing for whatever reason, according to Dr. Dietert, "we may not be getting some of the nutrients we need because they are not broken down to useful by-products." Microbes also break down environmental chemicals. Again, if they are missing, according to

Dr. Dietert, "We may not have some of the protection, and we may not have the seamless boundary that those microbes provide for us as we interact with our environment."

Like every single human being that ever walked this Earth, you are a whole human superorganism. Your body is a perfect ecosystem, filled with trillions of bacteria working in perfect harmony with your trillions of human cells.

At least that's what it's like if you're healthy.

Are we getting sicker?

You don't have to be a scientist or a doctor to notice that people living within industrialized nations are getting sicker. It's reached such a level that in *Missing Microbes*, Dr. Blaser described these populations as facing "a rising epidemic of plagues."

According to Dr. Blaser, asthma rates have gone up four- or fivefold in the United States since World War II. Food allergies are skyrocketing. As just one example, in the United States, the Centers for Disease Control and Prevention (CDC) released a study in 2013 revealing that food allergies among children had increased approximately 50 percent between 1997 and 2011.[9]

The incidence of juvenile diabetes, also known as type 1 diabetes, is also rising fast. In a study published in December 2015, researchers found that the number of children in the United States living with type 1 diabetes had increased by 60 percent between 2002 and 2013. Unlike type 2 diabetes, juvenile diabetes has nothing to do with obesity. Instead, it is an autoimmune condition in which the immune system attacks the cells that produce insulin, the hormone that regulates blood sugar.[10]

Celiac disease is another autoimmune condition that is on the rise. Gluten in the diet of someone with celiac disease triggers an immune response, whereby the immune system attacks the cells of the small intestine. In the United States "celiac disease has gone up four-fold since 1950," according to Dr. Blaser. Latest estimates suggest the condition now affects 1 in 100 people worldwide.[11]

Obesity is another upward trend. At the time of writing, the latest statistics from the US government state:[12]

- Two-thirds of US adults are considered to be overweight or obese
- One-third of US adults are considered to be obese
- One in twenty US adults are considered to be extremely obese
- One-third of US children aged between six and nineteen are considered to be overweight or obese
- One in six US children aged between six and nineteen are considered to be obese

Of course, the growing obesity problem is not just confined to the United States. It's happening in many other industrialized and developing nations as well.

Dr. Blaser has also noted a rise in gastroesophageal reflux disease, called GORD in the UK (GERD in the United States). This is described on the NHS Choices website as "a common condition where acid from the stomach leaks out of the stomach and up into the esophagus (gullet)."[13] The incidence of this condition has gone up dramatically since the 1930s, when it first appeared in the medical literature. According to Dr. Blaser, "It is associated with a premalignant disease called Barrett's Esophagus, first discovered in England by Sir Norman Barrett. This disease is a precursor to a certain cancer of the esophagus. And that cancer, adenocarcinoma of the esophagus, is the most rapidly increasing cancer in the United States and many developed countries. It's gone up six-fold over the last 30 years."

Cases of complex neurodevelopmental conditions, such as autistic spectrum disorder (ASD), are also increasing. Dr. Blaser says, "Autism is rising dramatically. There have been differences in diagnosis, and medical diagnosis is an imprecise field, but we think that autism has gone up at least four-fold since 1950. Some people estimate it's higher."

So, what's driving this rise in disease? Is it multi-causal, or is it just possible there's one reason why our population is getting sicker?

Dr. Blaser expresses a theory in *Missing Microbes* that gives a simple explanation for this rising tide of disease. "Let's say you have ten diseases that are rising at the same time; each one might have a separate cause, or perhaps there's one thing that's causing all of them to rise. I think that one thing is a change in our microbiome. Our ancient microbiome that protected us against many diseases is degrading, and with that degradation, these diseases are being fuelled."

If each of us carries many thousands of different bacterial species in our bodies, and if that's the way it's always been up until recently, those of us living in industrialized nations have somehow lost some of the diversity of bacteria in our gut, which is supported by recent research studies. In other words we have fewer different types of bacterial species living within and on us than we used to have.

Dr. Dominguez-Bello reported in *Science Advances* in April 2015 that members of an isolated American Indian group in Venezuela have the most diverse microbiome ever discovered in humans.[14] Spotted from the air by a helicopter in 2008, the Yanomami community has lived as hunter-gatherers undisturbed for the past 11,000 years in a remote mountain region. Researchers collected mouth, fecal, and skin samples from thirty-four individuals, being very careful to have minimal contact with the Yanomami people themselves.

Dr. Dominguez-Bello and her colleagues found that members of the Yanomami community had about 50 percent more ecological diversity than the average American. The research team hypothesizes that populations start to lose bacteria in their human microbiome as the culture becomes more industrialized. According to Dr. Dominguez-Bello, "When we compare our microbiome in the West with those from isolated peoples living in the jungles of South America, we estimate that we have lost about a third of the diversity of our microbes."

It's too soon to know for sure what losing a third of our microbial diversity means for people living in industrialized nations; however, this research has shown that members of the Yanomami community have, according to Dr. Dominguez-Bello, "a much richer microbial community than we do. And they are healthy."

If Dr. Blaser's theory in *Missing Microbes* is correct, we could be getting sicker because we have lost the diversity of bacterial species that have protected us for countless generations. As he says, "It's pretty clear that our diversity is going down. And that, to an ecologist, is dangerous because it's the diversity that protects us."

If there is evidence to suggest that we are losing the diversity of microorganisms in our bodies, the question is why. According to Dr. Blaser, "I would say part of it is modern life. Nothing prepared our microbiome for clean water, smaller families, Cesarean sections, and antibacterial substances everywhere. And of course 70 years of antibiotics."

How do antibiotics affect microbial diversity?

Over the past few decades, there has been a massive increase in the use of antibiotics both within medicine and within farming.

With regard to the use of antibiotics in medicine, a recent study from the Children's Hospital of Philadelphia looked at the health records, collected between 2001 and 2013, of 65,000 US children. Lead author Dr. L. Charles Bailey found that 69 percent of the children had been exposed to antibiotics before the age of two, with a mean average of 2.3 doses of antibiotics per child by that age.[15]

In many countries it's become common farming practice to feed antimicrobial drugs (which includes antibiotics, antifungal, and antiparasitical drugs) to animals destined for the food chain. Farmers use these drugs—in staggering volume—to prevent infection and to make the animals grow faster. A recent study conservatively estimated that, globally, animals on farms consumed 63,151 tons of antibiotics in 2010.[16]

A report published in the UK in December 2015 found that the amount of antimicrobials used in food production on a global scale is at least the same as their use in humans. In some places, animal use is much higher. For example, in the United States more than 70 percent of the antibiotics intended for human medical use are in fact given to animals.[17] According to the Soil Association, antibiotic use on UK farms is on the increase, rising 18 percent between 2000 and 2010. Now nearly 45 percent of all UK antibiotics are used in farming.[18]

Still on the subject of widespread and increasing antimicrobial drug use in farming, according to a recent study published in *Proceedings of the National Academy of Sciences*, the massive growth of factory farming around the world is expected to increase antimicrobial consumption in livestock by 67 percent globally by 2030. The amount used in countries such as Brazil, Russia, India, China, and South Africa is expected to double in the next 15 years.[19]

How do antibiotics work?

Antibiotics are medicines that treat bacterial infections. They work either by killing bacteria outright or by inhibiting the growth of bacteria, thereby giving time for the body's immune system to fight the infection.[20]

Some antibiotics target specific harmful bacteria. These are called narrow-spectrum antibiotics. One example of a narrow-spectrum antibiotic is penicillin G, often prescribed for syphilis, meningitis, pneumonia, and lung abscesses, as well as septicemia in children. Other antibiotics are broad spectrum, which means they don't differentiate between species of bacteria—they kill the good and the bad bacteria; the beneficial bacteria as well as the pathogens. Tetracycline is one such broad-spectrum antibiotic and may be prescribed to treat pneumonia and other respiratory tract infections, acne and skin infections, and genital and urinary system infections.

What is the origin of antibiotics?

Before the discovery of antibiotics, human health was in crisis. Bacterial infections such as tuberculosis, cholera, and even the bubonic plague killed millions of people around the planet. According to Dr. Blaser, "In the nineteenth century, there were epidemics of whooping cough and scarlet fever, and a lot of children didn't survive childhood. We were at war with some of these terrible pathogens. Fortunately, in the twentieth century, medical science created and developed a new series of drugs that we call antibiotics, starting with Alexander Fleming's discovery of penicillin."

The first true antibiotic, penicillin, was discovered by accident by Scottish biologist Alexander Fleming in London in 1928.[21] The story goes that, returning from holiday, Fleming opened one of his petri dishes containing colonies of staphylococcus (the bacteria that cause boils, sore throats, and abscesses) to find something unusual about the dish. Most of the dish was covered with bacterial colonies, apart from one area where blue-green mold was growing. The circular area around the mold was clear, and the mold seemed to be secreting something that was stopping bacterial growth. Fleming grew a pure culture of the bacteria-inhibiting mold-juice, which Charles Thom later identified as *Penicillium notatum*, a species of fungus now called *Penicillium chrysogenum*. Interestingly, the name *Penicillium* actually came from the fungus's resemblance to a paintbrush—*penicillus* is the Latin word for "paintbrush."

Fleming's mold-juice could kill a wide range of harmful bacteria, including streptococcus, meningococcus, and the diphtheria caused by bacillus, but it proved to be very unstable and it was difficult to produce in large quantities. It was more than a decade later, in 1939, when Howard Walter Florey, Ernst Boris Chain, and their team at Oxford University turned penicillin into a lifesaving drug.

Fleming's discovery, and its subsequent development for medical use by Florey and Chain, is undoubtedly one of the greatest discoveries of the twentieth century, saving millions of lives. Fleming, Florey, and Chain were jointly awarded the Nobel Prize in Medicine in 1945.[22]

A number of US pharmaceutical companies began working on producing penicillin on an industrial scale during the 1940s. Their aim was to prevent injured soldiers dying from bacterial infections. However, they needed a productive strain that could produce a lot of mold. That strain was found in a single moldy cantaloupe melon from a fruit market in Peoria, Illinois.[23] That moldy melon revolutionized twentieth-century healthcare.

What are antibiotics today?

Antibiotics have become a mainstay of modern medicine, saving lives every day. They underpin so many medical processes for all manner of treatments and conditions: We need them for safe practice in general surgery and transplant surgery, in chemotherapy, in the treatment of some dangerous bacterial infections—to name a few. Professor Stefan Elbe believes that because we have had such easy access to antibiotics over the past few decades, "we have lived in quite a privileged period of human history."

However, that privilege comes at a price. We are increasingly using antibiotics, to the extent that we are misusing them in many industrialized nations. This misuse has led to the growing threat of antibiotic resistance, a hot topic of political and medical discussion right now.

Antibiotic resistance means that antibiotics that were once effective against certain strains of bacteria are no longer so. This means that a number of serious diseases are becoming resistant to antibiotic treatment. According to an opinion piece by Mike Turner in the *Guardian* newspaper in May 2014 (based on the World Health Organization report published the same year on global antimicrobial resistance), six diseases that were thought to be long gone in industrialized nations "could come back with a vengeance."[24] These six diseases are tuberculosis, gonorrhea, klebsiella, typhoid, syphilis, and diphtheria. According to the article, "already diseases that were treatable in the past, such as tuberculosis, are often fatal now, and others are moving in the same direction. And the really terrifying thing is that the problem is already with us: This is not science fiction but contemporary reality."

Are we knocking our inner ecosystem out of balance?

Sometimes antibiotics are essential and can often can be life-saving—for example, if you're unlucky enough to get a serious infection such as sepsis. Sometimes a doctor prescribes antibiotics to clear up infections that aren't life-threatening, but when they might be trusted to work. However, while antibiotics might successfully treat your bacterial infection, they might also cause you problems elsewhere in your body. They might, for example, upset the balance of your gut microbiome.

You might have seen this for yourself if you've ever had diarrhea after a course of antibiotics. Sometimes it takes a while for everything to settle down again. According to Dr. Neena Modi, "Well, that's our microbiome groaning and saying 'Hey, something's being done to me.'"

If you're a woman, you might have taken antibiotics for a few days and then have experienced a vaginal yeast infection, often known in the UK as vaginal thrush. According to the UK government NHS website, one-third of all women who take antibiotics develop vaginal thrush.[25] The course of antibiotics could have knocked your vaginal microbiome out of balance. Broad-spectrum antibiotics wipe out many species of bacteria all over your body, including some species of beneficial bacteria in your vagina. These bacteria are the ones needed to keep your body's yeast population in balance inside your vagina. When these bacteria are missing, the yeast multiplies until you have a yeast overgrowth (typically, of the fungus candida), which presents itself as thrush. Over-the-counter medicines should easily treat the yeast infection, and the NHS information website lists other treatment suggestions, including complementary and natural therapies, such as eating live yogurt (which is full of beneficial bacteria) to restore the natural balance.

A useful way to explain the importance of microbial diversity in our human bodies as adults is to think in terms of an analogy. Dr. Dietert uses the analogy of a diverse forest: "If you remove half of the types of tree in a very diverse forest, the forest doesn't stay the same for the rest of the trees that are there. There's extra space. The dynamics of the interactions are completely changed. The insects, the animals may be changed because of the change in their habitat, and the entire forest takes on a totally different flavor. In some cases, a very unpredictable flavor."

Removing just one species is enough to affect the interactions between all the remaining species within an ecosystem. The habitat changes; some species start dying off; other species start taking over, resulting in overgrowth. The balance shifts; the ecosystem changes; everything is thrown out of balance.

Using Dr. Dietert's example of a diverse forest, if you remove all leafy oak trees from a dense forest, more sunlight can hit the forest floor. The plants that like the cool shade die out and other sunlight-loving plants move in. The insects that once fed on the shady forest-floor plants fly off to find food elsewhere, with the forest animals that eat those insects not far behind. Different insects appear, followed by animals that eat those kind of insects. Maybe there's one really dominant insect that really likes these new conditions. That insect reproduces at a phenomenal rate and suddenly you have swarms of that one insect species—too many of them for the sunlight-loving plants to sustain. The plants die, and the whole ecosystem begins to change again. All because we removed just one species of tree.

In this example it might be that the leafy oak tree is what Dr. Blaser calls a keystone species. These are species that are few in number but have a powerful effect on all the other organisms in the ecosystem. According to Dr. Blaser's theory, our bodies could host a number of keystone species. If you disturb the population of any these keystone species as a result of, for example, taking antibiotics, there could be consequences for the health of the whole human superorganism.

This could be happening right now in your body. Any action you take could disturb your inner microbial ecosystem. As Dr. Blaser explains, if something disturbs the fragile equilibrium, this could potentially make the whole ecosystem more susceptible to a wide-scale attack. "As one after another of our microbes becomes extinct, we become more vulnerable to diseases like obesity, diabetes, and asthma. That means shorter lives. It means lives with more susceptibility to the big pandemics. And those are the things that are most deadly. So there's a lot of risk out there." We'll explore the implications of this idea later, in chapter 8.

What lies before us might be a potentially very scary public health crisis. But there's hope. Recognizing the problem is the first step to finding a solution. As Dr. Blaser said in his interview with us, "We have the tools and if we had the way, we could fix it."

The good news is that we already have "the way," as Dr. Blaser describes it. "The way" starts with raising awareness of the microscopic events happening at the very start of life.

And it all starts with birth.

Here's a summary of the main points we've covered in this chapter:

1. The first life-forms on Earth were microbes, which successfully colonized the whole planet. All other life-forms since have evolved with microbes, including all humans.
2. Humans are superorganisms—our bodies comprise trillions of human cells and trillions of microbes.
3. The microbes that live on and in our bodies and call us home (bacteria, viruses, archaea, fungi, and so on) are collectively known as the human microbiome.
4. Bacteria live on our skin and inside us; in our ears, noses, mouths, lungs, and crucially, because of a connection with our brain, inside your gut. If you're a woman, they live inside your vagina.
5. The presence of bacteria helps keep your body functioning properly and helps protect it from disease.
6. Aspects of living in industrialized countries, including diet, lifestyle, antibiotics, antibacterial products, and rising rates of C-section, have reduced the diversity

of microbial species in our bodies—estimates are by up to a third.

7. According to Dr. Martin Blaser, this loss of microbial diversity could be "fueling" the rise of many common noninfectious conditions, including allergies, asthma, diabetes, autoimmune disorders, obesity, some mental health conditions, and even some cancers.

WHAT IS THE MICROBIOME?

TWO

What Do Bacteria Have to Do with Birth?

Without birth there is no life. Birth is the moment you emerge into the world; the moment you take your first breath; the moment you meet your parents and the moment they meet you. And it happens only once in your life.

For Professor Hannah Dahlen, childbirth is so much more than the arrival of a new human being into the world. "Birth is a neurohormonal, mechanical, immunological, microbiological, social, psychological, emotional, cultural, and spiritual event."

For neonatologist Dr. Neena Modi, "The moment of birth is accompanied by such profound experiences on every single level; at a cellular level, at a metabolic level, pain responses, stress responses. You name it, birth is a very profound moment indeed."

What most people don't realize is that there are actually two events happening during childbirth.

First, there's the amazing event of the birth itself, the arrival of a new human into the world. As parents, we remember every single second of our daughter's birth. And it seems that other parents we talk to remember their children's births in great detail, too. Our own parents tell us they also remember everything about our births and those of our siblings (Toni is one of five children, Alex one of four). It seems that for many parents (if not all of them), memories of a child's birth last a lifetime.

So, what else is happening during childbirth? What's the second thing? It's something that we weren't aware of when our daughter was born—and now that we know about it and how crucial it is, we wish we'd had some inkling, because this knowledge would have informed the choices we made at that time. This second event is microscopic—invisible to the naked eye—and yet it could play a significant part in determining a child's health for the rest of their life. Indeed, this event could already have determined our daughter's health for the next 80 or more years and increased the risk of her developing serious health conditions throughout her life. At the time of her birth, though, we were completely oblivious to it.

So what is it? The second critical event happening during childbirth that could have lifelong health consequences for a child is . . .

. . . the main seeding of the baby's gut microbiome.

Other factors, such as genetics, the mother's health and diet, and the mode of infant feeding may also play their part in the establishment of the baby's gut microbiome, but the latest research indicates that the way in which we are born could be critical. The means by which we enter the world could significantly influence the composition and diversity of our gut microbiome. Crucially, it could significantly impact the subsequent development of our immune and metabolic systems, from infancy throughout life.

In other words what happens at birth has consequences that last a lifetime.

As we're about to discover in the next section, a baby might receive a small degree of prenatal exposure to microbes during pregnancy. Indeed, some of these microbes from the womb, placenta, or amniotic sac could even reach the baby's gut before birth; then many more microbes arrive with the main seeding event happening during childbirth. As humans, we've evolved so that specific bacterial species are the first to colonize the gut microbiome. These microorganisms set up the gut "colonization party." But before we get too far ahead of ourselves, let's take a step back to look at how the mother's microbiome prepares for birth.

What happens microscopically during pregnancy?

Babies develop in a near-sterile environment. It's not completely sterile; recent research indicates that babies may receive some prenatal exposure to bacteria while in the womb. Scientists have detected that small colonies of bacteria may be present in the womb, in the placenta, in the amniotic fluid, and perhaps even in the fetus's developing intestines.[1]

At the time of writing, scientists have discovered that at least a third of pregnant women have bacteria present in their placentas.[2] Research suggests there's a possible link between the presence of bacteria in the placenta and the mother having had an infection earlier in pregnancy, such as a urinary tract infection during the first trimester. There's also a link between the presence of bacteria in the placenta and preterm birth.[3]

Research also indicates that if bacteria are present in the placenta, the microbial profile of the placental microbiome resembles the microbial profile of the oral (mouth) microbiome. In other words, there could be a connection between the colony of bacteria present in the mother's mouth and the colony of bacteria present in the mother's placenta. But how could bacteria from the mother's mouth end up in the placenta?[4]

As described in a December 2015 article in *TIME* magazine titled "Babies in the Womb Aren't So Sterile After All," scientists suggest it's possible that bacteria might travel from the mother's mouth, down through her bloodstream, and end up in the placenta. Or it might be that certain microbes travel from the mother's vagina into the womb. With more research, hopefully it won't be too long before we will have more definitive answers. If there is a connection between the placental microbiome and preterm birth, knowing why the bacteria are present and how they reached the placenta could potentially present a pathway to prevent or help reduce the number of preterm births in the future.[5]

To sum up, the latest research indicates that at least in some pregnancies, the fetus could receive some prenatal exposure to small colonies of microbes in the uterus. This could mean a small degree of prenatal seeding of the fetus's gut microbiome occurs during pregnancy. With this in mind, the mother's diet, lifestyle, and perhaps even her dental hygiene during pregnancy could influence the baby's gut microbiome even before the baby is born.

Whatever the extent of prenatal exposure, the actual main seeding event for the baby's gut microbiome happens, as described by Dr. Rodney Dietert, in the "narrow window that surrounds birth." This main seeding event occurring in and around childbirth involves the colonization of significant quantities of bacteria to establish the baby's gut microbiome.

In the months leading up to the birth, the pregnant mother's whole microbiome changes in preparation for the transfer of microbes from mother to baby that happens during the birth process. The biggest changes occur in the mother's vaginal and intestinal microbiomes because these will play an important part in the "vertical microbial transmission to the newborn during vaginal delivery," to quote from Dr. Maria Gloria Dominguez-Bello's 2015 paper "The Infant Microbiome Development: Mom Matters."[6]

In her interview with us, Dr. Dominguez-Bello describes exactly how a woman's microbial population changes as she progresses through her pregnancy: "Her microbes start changing; her intestine changes in terms of the microbial communities, and the vaginal bacteria also change. In the vagina, we know that she will, especially toward the third trimester, drastically increase the amounts of lactobacilli, the bacteria that will colonize the baby."

Lactobacilli bacteria are often described as "friendly bacteria." They are found in yogurt and other fermented foods, as well as in oral cavities, the gastrointestinal tract, and the vagina.

A little biology lesson might be handy at this point. Lactobacilli are part of the lactic acid bacteria group, whose members convert lactose and other sugars into lactic acid, which helps provide energy. Lactose is the principal carbohydrate of breast milk, giving it a strong connection with lactobacilli. If lactobacilli bacteria meet breast milk, they break down the lactose in the milk to produce energy for the baby. That could be why the mother's vaginal microbiome shifts late in pregnancy so that there are more lactobacilli bacteria. It's a perfect evolutionary adaptation in preparation for birth and subsequent breastfeeding. Not only does the expectant mother's vaginal microbiome change so there's a higher proportion of lactobacilli bacteria, but the diversity of microbial species actually decreases so that there are fewer other bacterial species present. So, lactobacilli species (*L. vaginalis*, *L. crispatus*, *L. jensenii*, and *L. gasseri*) increase inside the mother's vagina. So much so that they crowd out the other bacterial species, limiting bacterial diversity overall.

As Dr. Dominguez-Bello notes in her "Mom Matters" paper, having so many lactobacilli present in the vagina late in pregnancy helps maintain a low pH (low alkalinity/high acidity), "thereby limiting bacterial diversity and preventing bacteria from ascending to the uterus, where they can infect the amniotic fluid, placenta, and fetus."[7]

Higher acidity helps prevent infection spreading to the amniotic fluid, placenta, and baby. With birth involving lactobacilli as the "seed" and with breastfeeding involving lactobacilli as the "feed," it would appear that lactobacilli bacteria are critical to both ends of the seed-and-feed process. We'll cover more on this in the next chapter. Suffice to say, lactobacilli is one of the bacterial species names to remember when talking about birth *and* breastfeeding.

During pregnancy, the mother's metabolism changes. As she is carrying a growing fetus, she needs more energy. The latest science indicates that it could be the pregnant mother's microbes

41

that help the mother extract more energy and nutrients from her food. According to Dr. Dominguez-Bello, "In the mother's gut, the bacterial components will have a structure that will improve the energy extraction from the diet. Mothers almost have a metabolic syndrome because they need to extract energy so badly. The baby is such a strain on the mother's energy that bacteria help to prepare her body for this state."[8]

During vaginal birth the baby is likely to come into contact with the mother's fecal matter. This fecal matter contains many of the mother's intestinal gut microbes, potentially including another important type of "beneficial" bacteria, intestinal bifidobacteria (another member of the lactic acid bacteria group).[9] We'll return to this subject later in this chapter, but suffice to say now that this contact with the mother's fecal matter during the birth process is likely to be a really good thing. In terms of energy release: What if we have evolved so that newborns could benefit from having immediate access to those high-energy-yielding microbes found in the mother's fecal matter? What if that exposure to the mother's fecal matter helps give the baby an energy boost for the critical first few hours of life? After months cocooned inside the mother's womb, you can imagine that those first few hours after birth are pretty energy-sapping for a newborn. The baby sometimes has to squirm up the mother's abdomen and chest to find the nipple, then to quickly master how to latch on, suck, and swallow. On top of that there are all the incredible new experiences to take in and adjust to, with a wealth of new smells, sights, sounds, tastes, and touch sensations. Those high-energy-yielding microbes would come in really handy, especially if the newborn is mostly awake for the few hours after birth.

As Dr. Dominguez-Bello writes in her "Mom Matters" paper, there are still many questions to answer regarding exactly why and how the mother's microbiome changes during pregnancy. However, "gestational changes in the vaginal and fecal microbiota are likely to be part of an adaptive response to protect and promote the health of the fetus and provide the newborn with a specific microbial inoculum at birth, before exposure to other environmental microbes."[10]

It could be that all these microscopic changes in the expectant mother's body are deliberate, that they are a result of many evolutionary adaptations that have happened across millennia. It could be that these changes should happen at a certain time, in a certain way, to perfectly prepare the mother's body for the transfer of the right types of microbe at the right time to her baby. The microbial changes create a special cocktail of exactly the right mix of microbes in the mother, all ready and waiting for the big moment to arrive when the cocktail is transferred to her baby.

With all this in mind, the bacterial populations in the mother's body would ideally be in the best possible condition, ready for transfer to the baby. Perhaps extra-special attention could be given to an expectant mother's diet and lifestyle during pregnancy, in order to best prepare the mother's gut and vaginal microbiome for birth. Perhaps extra-special attention could also be given as to whether antibiotics are administered during pregnancy, especially as they could significantly affect the mother's—and through the seeding process the baby's—gut microbiome.

How is a baby's microbiome seeded at birth?

Before answering this question, we first have to explain the physiology of natural birth. Scientists still don't fully understand the exact process that triggers spontaneous labor for full-term babies, but it's believed to involve the maturity of the fetus in the womb and the interaction of hormones and physical changes between mother and baby.

From just after the middle of a pregnancy, an expectant woman might start to feel mild practice contractions known as Braxton-Hicks contractions. These tend to be infrequent, irregular, and unpredictable contractions that last less than a minute. They often stop if the woman changes activity or position, say from walking to sitting still. Braxton-Hicks contractions combined with the cervix getting shorter and more stretchy, plus changes in hormone levels, may all combine to ripen the cervix for labor. Toward the very end of her pregnancy, say between 38 and 42 weeks, the expectant mother might experience some slightly stronger contractions that midwives call pre-labor "tightenings." The sensation of tightening in the abdomen or bump might be accompanied by mild abdominal pain, similar to period pain, and some lower-back pain. The pre-labor tightenings may come rhythmically in late pregnancy, perhaps every ten or twenty minutes. For some women, around this time, the mucus plug that has helped maintain the near-sterile environment for gestation comes away from the neck of the womb—this is called a show.

At some point contractions might start to become more intense—they last longer and become stronger, more frequent, and more painful. This is the crossover point, when pre-labor tightenings become early labor. Once the contractions come three or four times in ten minutes, and the contractions are strong and regular, a woman is in "active labor"—when the baby starts to begin its epic journey from the womb to the outside world.

Throughout her labor, the mother will release several naturally occurring hormones. One of those hormones is oxytocin, the so-called love hormone, which stimulates stronger contractions, forcing the baby to bear down on the cervix, softening and widening it so that it opens to a diameter of about 10 cm. Once the woman's cervix has reached full dilation, the baby passes through it and into the birth canal, making its final journey toward the opening of the vagina. This is the most painful and intense phase of active labor.

Usually at some point during labor, the woman's waters will break. This happens when the amniotic sac—the fluid safe house the baby has been living in while in the womb—splits open. Very occasionally waters break before labor actually starts—the first thing a woman feels could be a gentle trickle or sudden gush of water running down her legs. (In the movies, the waters always seem to break at an inopportune time—in a supermarket or at a public event. But in reality it can happen at any time, even while the mother is at complete rest, such as when she's sleeping or doing something completely innocuous such as washing up!)

Most of the time, though, the waters break either during the first stage of labor, when the contractions begin to intensify, increasing the pressure on the amniotic sac so that it pops, or during the second, pushing stage, when the baby is in the birth canal. Rarely, a baby may be born with the amniotic sac intact, which is known as being born en-caul.

The moment that the waters break (if they do) is the critical moment for the main seeding of the baby's microbiome. As soon as the membranes of the amniotic sac rupture, the baby is suddenly exposed to an influx of bacteria. In the birth canal, without the protection of the amniotic sac, the baby becomes coated in the mother's vaginal microbes, which the newborn skin soaks up like a sponge. The microbes enter the baby's eyes, ears, and nose, and make their way into the baby's mouth. According to Dr. Martin Blaser, "The baby is covered with the bacteria; the baby is coated, is swallowing the bacteria. That's the baby's first introduction to the world of bacteria, the world that we all live in. That's the founding microbiome for that baby."

Some of the microbes that enter the baby's mouth stay in the baby's mouth, populating it in particular with lactobacilli. These bacteria will prove crucial later, when the baby starts to breastfeed. Not all the vaginal microbes stay in the mouth, however—the baby swallows some (including lactobacilli), sending the mother's vaginal microbes down through the baby's gastrointestinal tract and into the gut. Here, they start to form the foundation of the baby's very own gut microbiome.

The mother's vaginal microbiome includes a high percentage of lactobacilli bacteria, which, as we've already described, help extract energy from the lactose in breast milk. So by being among the first species in the gut, the lactobacilli are all ready and waiting for the breast milk to arrive. But not just that. According to Dr. Blaser, lactobacilli also have another special weapon: "These species are also armed with their own antibiotics that inhibit competing and possibly more dangerous bacteria from colonizing the newborn's gut."[11]

What this means is that lactobacilli bacteria have their own kind of antibiotics that help stop other potentially harmful bacteria getting a foothold in the same space.

There's one last amazing thing associated with the first arrivals in the baby's gut—they help train the immune system. This demonstrates the brilliance of the whole process: First, the microbes colonize the baby's gut microbiome; second, they digest lactose to produce energy; third, they play an active part in preventing harmful pathogens taking hold in the baby's gut; and finally they help initiate the training of the immune system.

Back to the baby's epic journey through the birth canal. Professor Dahlen describes it perfectly: "In the birth canal, the baby kicks up to use the top of the mother's uterus as a diving board. It is constantly moving and assisting. It wriggles its head in a series of movements through the pelvis, which helps it find the widest diameter, almost doing a 180-degree turn. As the baby comes down, it is squeezed. The compression releases a huge cascade of hormones and helps the baby empty its lungs, helping the baby to get ready to be born."

Once the baby has made it through the main bones of the pelvis, it's time for the actual moment of birth. The baby comes up under the mother's pubic bone and begins to crown—when the top of the head becomes visible to the outside world. With further contractions and a reflex to push, in a straightforward vaginal birth, the baby's head, then the first shoulder, wriggle out. Once the second shoulder wriggles free, the rest of the baby emerges, and so the baby is born.

The length of time the baby is exposed to its mother's vaginal microbes depends upon when exactly the amniotic sac splits open (the moment the main seeding event begins) and exactly how long the baby spends in the birth canal from this moment. The birth canal is commonly defined as the channel from the cervix to the vulva. For some relatively quick births, if the waters break before or during the second pushing stage, exposure time to the mother's vaginal microbes in the birth canal could be a matter of minutes. For slower births, particularly with a longer second, pushing stage, the exposure time could be hours.

As soon as the baby emerges into the outside world, he or she receives more microbes: from the air, and from everything and everyone that touches the baby; from being handled, from being kissed, from contact with the mother's chest.

If the waters broke during labor, the gush of fluid is likely to have spread vaginal juices, packed with microbes, far and wide, perhaps even coating the mother's inner thighs and abdomen. If, at birth, the baby "lands" on the mother's thighs or abdomen, lactobacilli and other vaginal microbes immediately soak into the baby's skin, helping to colonize the baby's skin microbiome.

Even if the mother didn't defecate during the birth itself (which is a very normal event), if the baby was facing downward on the way out of the vagina, his or her face might have brushed up against the mother's anus. Fecal matter contains microbes from the mother's gut and is likely to include colonies of bifidobacteria, which are the aforementioned "beneficial" bacteria. Bifidobacteria not only help break down the undigested sugars in breast milk, but also coat the lining of the infant's intestine, plus their presence inhibits the growth of pathogens, and they also may play a role in the immune system. If the baby's mouth comes into contact with the mother's fecal matter, then some of these other "beneficial" microbes will be swallowed, to join in the microbial "colonization party" just beginning in the baby's gut.

Myriad different microbes, from various sources, arrive on and in the newborn baby in a very short space of time, quickly multiplying to colonize the baby's microbiome. This is a once-in-a-lifetime event. The microscopic exposures happening right at that moment, in the time during and immediately after birth, help lay the foundations for lifelong health and immunity.

Dr. Dietert believes childbirth should be viewed as the single most pivotal moment in a child's life for determining lifelong health. This is the premise for his groundbreaking "completed self hypothesis,"[12] which draws on and extends two hypotheses, the "fetal programming hypothesis"[13] developed by the British academic Professor David Barker and also the "hygiene hypothesis."[14]

The Barker fetal programming hypothesis proposes that the baby's environment in the womb and early life, influenced by the mother's nutrition, helps determine the baby's health in later life. The hygiene hypothesis proposes that early microbial exposure can be beneficial to long-term health. A large number of studies have shown that growing up on a farm helps protect a child against allergies.[15] Perhaps, then, city children are at increased risk of allergic diseases because they are "too clean"; they might not have the same microbial exposures as those born and raised on a farm.

Dr. Dietert goes one step further. He believes an infant exposed to the right microbes at the right time, starting from birth, achieves the healthiest human state throughout life. "That is how we are designed, that is what is intended, that is our healthiest state. That's the way it should be. And that needs to happen early in life so that the immune system matures effectively."

According to Dr. Dietert, if you miss out on getting the right microbial exposure at birth, you miss out on the full process of educating your immune system: "The immune system is not trained appropriately so it starts to react to anything and everything, and winds up destroying the tissues in the body." (See chapter 5.)

Microbes received in early life also play an important role in the baby's metabolic system. According to Dr. Dietert, "Those microbes are also really important in breaking down food and protecting us from environmental toxicants. If they are missing, we may not be getting some of the nutrients we need because they are not broken down to useful by-products. We may not have some of the protection microbes provide as we interact with our environment."

Dr. Dietert believes that a baby's full set of microbes obtained as a result of vaginal birth enables the child to be a "complete" human superorganism. The optimal seeding of the baby's microbiome that happens with vaginal childbirth helps a person become who they are supposed to be (microbially speaking), with exactly the right microbes they are supposed to have.

What happens if a baby is born en-caul?

An en-caul birth is one where a child is born inside the entire amniotic sac. The sac balloons out at birth with the child still inside the thin filmy membrane, the amnion. This is different from being "born with the caul," which is where the thin filmy membrane covers just the baby's head and face. It is a relatively rare birth event, occurring in around 1 in 80,000 births.[16] As the baby is attached to the placenta throughout birth, he or she continues to receive oxygen (and nutrients) from the mother, and until his or her face emerges from the amniotic sac and hits the air, the lungs have no need to function. In many cultures, being born with the caul is considered very lucky.

However, what about the vaginal microbes? How can a baby who is born with membranes still covering his or her face, or even his or her whole body, receive a dose of the mother's vaginal and intestinal bacteria?

We don't yet know the answer to these questions—at the time of writing, there have been very few, if any, high-quality research studies to find out if a baby born with the caul misses out on colonization by the mother's vaginal or intestinal microbes. The absence of research on this topic is probably partly because a caul birth is so rare and partly because this area of microbiome research is still so new.

Right now, all we have is supposition and speculation. We might speculate, for example, that a baby born partly or wholly with the caul doesn't receive a full complement of the mother's microbes during labor itself, although once the amniotic sac is ruptured after birth (usually by a doctor or midwife), we can surmise that, as we'd expect, the baby starts receiving microbes straightaway—through skin-to-skin contact, through contact with the mother's vaginal fluid or fecal matter spread over the mother's thighs and abdomen (see page 47), and from the air and environment into which he or she is born. In effect once the doctor or midwife has removed the caul (usually via a small incision in the membrane across the baby's nostrils that allows the midwife or doctor to peel the rest of the caul away from the skin), the baby receives microbes in the same way as any other. Whether or not there are microbial benefits for remaining within the amniotic fluid remains to be seen.

What about a water birth?

At Q&As following screenings of our film *Microbirth*, there is usually at least one question about water birth. Many expectant parents, doctors, midwives, doulas, and birth educators want to know how water birth affects the seeding of the baby's microbiome. The simple answer (and potentially most exciting answer for scientists) is that we just don't know. At the time of writing, there have been very few high-quality, large-scale studies on the microbial impact of babies born in water. We simply don't know for sure if being immersed in water during labor and birth has an impact on the seeding of a baby's gut microbiome, and (if it does) what that impact might be. According to Dr. Dietert, "Not much if anything has been published on this specific factor. So far, most of the research has focused just on vaginal birth versus Cesarean section delivery."

As we've explored previously, in a vaginal birth the main seeding of the baby's gut microbiome starts to happen while the baby is in the birth canal (after the amniotic sac has burst open), then continues with contact with the mother's fecal matter, then more exposures through being touched, kissed, and handled after birth. If the mother is laboring outside a birth pool when her waters break, the main seeding process begins at that moment, the same as with any other vaginal birth. If the laboring mother subsequently gets into the birth pool, some of her vaginal or intestinal microbes may be washed away; or maybe the pool water makes no difference at all—we can only speculate.

Alternatively, if the waters break while the mother is laboring in the birth pool, or if the mother actually gives birth while in the pool, the mother's vaginal and intestinal microbes may spread around and within the birth pool water. At the moment of birth, water containing these microbes could still enter the baby's eyes, ears, and nose, and the baby may swallow some to colonize his or her gut. Other microbes may coat and colonize the baby's skin microbiome. Once the baby is out of the water, the colonization process continues as the child is exposed to microbes from touch and from the air, just as with any other birth.

A key question is whether chlorine or other chemical detergents commonly used to clean water birth pools between births might affect the seeding process. Indeed, if the birth pool is filled with chlorinated drinking water (chlorine is a highly efficient disinfectant often added to public water supplies to kill harmful pathogens), perhaps the chlorinated water itself might also affect the seeding process. Altogether, only future research can tell.

What are the microbial benefits of vernix?

Vernix is the white, waxy substance with which babies are coated when they are born. In *Missing Microbes*, Dr. Blaser wonders about the benefits of vernix for the baby's skin microbiome: "While no one yet has studied this in detail, my hunch is that vernix serves to attract particularly beneficial bacteria and repel potential pathogens."[17]

Until we have scientific evidence, we won't know for sure, but Dr. Blaser's supposition makes us question the practice of rubbing off vernix within minutes of a baby's birth in hospitals all around the world.

What is the impact
for babies born before 38 weeks?

A pregnancy is considered full term if the baby is born after 38 weeks' gestation. We don't yet have a full picture of the potential impact of preterm birth on the development of the baby's microbiome.

While preterm babies born vaginally make their way into the world through the birth canal, the prematurity of birth may have an impact on the level of microbial exposure for that baby. We simply have to wait for more research for more definitive answers.

How does the baby's microbiome develop *after* birth?

As previously discussed, the first microbes to arrive in the baby's gut during vaginal birth are likely to come from the birth canal and hence could contain a high proportion of lactobacilli bacteria. To get a bit technical, these first arrivals are "facultative anaerobes," which means they can grow with or without oxygen. These facultative anaerobic microbes use up all the oxygen in a baby's gut, thus they help pave the way and create the ideal environment for colonization by populations of "obligate anaerobes." These are species of bacteria that only grow when oxygen is not present; examples include bacteroides, clostridium, and bifidobacterium.[18] Species of bifidobacteria will digest the indigestible sugars in breast milk (which will be discussed in the next chapter), growing and colonizing especially quickly, thereby making it difficult for unwanted pathogens to take hold. What all this could mean is that bacteria are perhaps meant to arrive in the baby's gut in a certain sequence. With vaginal birth, lactobacilli might arrive first to set up the "colonization party." They use up all the oxygen, enabling strict anaerobes like bifidobacterium to thrive.

Bifidobacteria are one of the "colonization party superstars"; they come from the mother's intestinal microbiome (from the baby's contact with the mother's fecal matter) and from breast milk. They multiply quickly, eventually becoming the dominant species in a newborn's gut, crowding out harmful pathogens. The bifidobacteria can either attach directly to the intestinal walls or become trapped within the mucous layer of the gut lining. Then in the coming days, weeks, and months, other bacteria arrive and continue the colonization process.[19]

The microbial profile is in flux for the first 2 or 3 years of life, but scientists believe it commonly stabilizes sometime between the ages of two and three (although this might depend on when breast-feeding stops). Once the infant microbiome reaches its equilibrium, it begins to resemble that of an adult—this means it is quite similar to an adult's in terms of composition and diversity. The latest science indicates that, once established, approximately 60 to 70 percent of the composition of the gut microbiome remains stable from childhood to old age. The remaining 30 to 40 percent may become altered as a result of, among other things, changes in diet or lifestyle, levels of stress and physical activity, bacterial infection, antibiotics, or surgery.[20]

In effect it seems that by the age of three, we have a baseline state of health. So, while throughout most of a person's life the precise nature of the microbiome might change from day to day or month to month in response to environmental or lifestyle factors, fundamentally it will most likely return to 60 to 70 percent of its baseline state. Until, that is, we reach our sixties. At around this time in our lives, age causes a shift in our physiology. This, perhaps coupled with a change in diet and lifestyle and an increased use of pharmaceutical drugs, means that the microbiome enters a period of flux—it is more unstable and often less diverse.[21]

Research indicates that in order to have a perfectly functioning microbiome, the critical period for microbial exposure is during those early years, before the microbiome stabilizes. This is the period when the immune system develops (see chapter 5). According to Professor Anita Kozyrskyj, "With respect to the development of the infant gut microbiota, we're learning it's a gradual development. You start with a few species then gradually over time, the pattern becomes similar to that of the adult. And during this time, the microbiota has a purpose. And that is in training the development of our immune system. This is a very good example of the interaction of the microbiota with the host."

Of course, those early years are also a key time in a child's cognitive, social, and physical development. According to Dr. Blaser, "Our concept of health is that there is a developmental cycle. Whether we're talking about growth, metabolism, immunity, or cognition, there are steps to development. Our hypothesis is that the microbes are part of this development. Our development and their composition are choreographed so that we're developing together. That's normal, that's healthy. When we interfere with that, then there might be consequences."

Now that we've explored what happens with the foundation of the baby's microbiome during childbirth, let's see how those processes are beautifully complemented by the microscopic events happening during breastfeeding. As we're about to discover, the whole seed-and-feed process happening during birth and immediately afterward is a perfectly exquisite natural system. It is a system that, if it is interfered with, could have health consequences that last a lifetime.

Here's a summary of the main points we've covered in this chapter:

1. During pregnancy, the mother's vaginal microbiome changes so that there are more species of lactobacilli bacteria.
2. During pregnancy, babies develop in a near-sterile environment in the womb. The latest science indicates that babies may receive some prenatal exposure to microbes from the placenta. Other microbes have been found in the womb and amniotic fluid in some pregnant women.
3. During birth, whatever the extent of the prenatal exposure, the main seeding event for the baby's gut microbiome happens in the narrow time frame that surrounds birth.

4. During birth, as soon as the amniotic sac splits open (the waters breaking), an influx of microbes floods over and coats the baby, entering the baby's eyes, ears, and nose, while the baby swallows some through the mouth.
5. With vaginal birth, the main seeding event happens through the massive bacterial colonization upon exposure to vaginal, fecal, and skin microbes.
6. During birth, the baby might also be exposed to the mother's intestinal microbes via contact with the mother's fecal matter—a very normal and potentially beneficial thing!
7. Once born, the baby acquires more microbes—from the air, from being touched, and from being fed. These microbes all join the "colonization party" in the baby's gut.

HOW THE MICROBIOME
IS SEEDED AT BIRTH

— THREE —

Breast Milk or Formula?

Once the baby is born, the next step in the optimal seed-and-feed process is immediate skin-to-skin contact, ideally with the mother. Or, if that's not possible, with the father or another close family member.

Imagine emerging from the safety, darkness, and relative peace, quiet, and warmth of the womb to the bright lights, the noise, and the general commotion of the outside world. It is no doubt a profound experience for a baby, who must be overwhelmed by new sounds, sights, smells, tastes, and feelings. Adjustment to life outside the womb might surely take a little while.

In the UK, midwives are encouraged to wait an hour before performing initial checks on the baby to allow parents and baby to spend a magical hour of skin-to-skin contact together and to give the baby some time to adjust to his or her new environment.

Professor Hannah Dahlen beautifully describes the physiological and emotional benefits of early skin-to-skin contact, particularly for the baby: "The natural habitat for a newborn baby is skin to skin with its mother. If this happens, the mother's high levels of cortisol released during labor start to dampen down, and the mother will release oxytocin. The baby starts to smell and search for the mother's breast. Skin-to-skin is so amazing—the mum will warm up if the baby's cold, being on the mother's skin will help regulate the baby's blood sugar. It regulates the baby's breathing, it reduces crying, it reduces stress. It basically brings the baby into the world in the gentlest way possible."

According to Lesley Page: "After birth, you'll often see a kind of classical bonding behavior: The mother gazes at her baby, stroking, holding, kissing, and talking to her newborn. It's an absolutely critical time. It's a moment when the mother and the father will see their baby's eyes, and the baby recognizes the human face. You have to allow time to unfold because they will never have that moment again. It's the first stage of falling in love, and it actually sets a template for future life."

Is there microbial transfer during skin-to-skin contact?

In chapter 2, we talked about how skin-to-skin contact immediately after birth enables the transfer of more bacteria from the mother's skin to the baby's skin, to help colonize the baby's skin microbiome. However, that's not the end of the story.

In 1987, Ann-Marie Widström and her team from the Karolinska Institute in Sweden described a phenomenon whereby a baby placed onto the mother's abdomen soon after birth may, if left to their own devices, climb up the mother's chest to find her nipple and begin to feed.[1] Widström calls this a "breast crawl," which she identified as having a number of different stages. She describes the stages beautifully in a paper she published in 2014: "We now know that the healthy newborn infant has an inborn sequential behavioral pattern during the first hours following birth if placed skin-to-skin on the mother's chest. Gradually, the reflexes come to life: The infant successively acquires sucking and rooting reflexes, fists the hand, brings the hand to its mouth at about half an hour after birth, and within one hour postpartum, finds the mother's breast and begins suckling."[2]

If the mother's abdomen up to the chest is coated in microbe-rich juices from the birth (see page 47), the baby's "breast crawl" is a brilliant opportunity for making the most of these juices, especially if the baby licks and sucks the skin during the climb up the chest. Licking is very common in newborns. According to Lesley Page: "Breastfeeding isn't immediate. The baby first licks and nuzzles." Licking and nuzzling, as part of the inborn sequential behavioral pattern described by Dr. Widström, would be an ideal opportunity for the baby to acquire more microbes not just to colonize his or her skin microbiome but also to colonize the baby's gut. This is pure supposition on our part, but perhaps acquiring additional microbes from instinctively licking and nuzzling could be one reason why a newborn doesn't initiate breastfeeding straightaway.

With all this in mind, common postnatal maternity practices, including checking, tagging, weighing, measuring, wiping off the vernix, and immediately cleaning the newborn baby, could all potentially interfere with the critical transfer of microbes from mother to baby during the crucial first couple of hours of life. Unless there's a medical emergency with either the mother or the baby post-birth, perhaps there's an argument for delaying some or all of these postnatal practices for at least a couple of hours, if not longer, so as not to disturb the optimal seeding of the baby's gut and skin microbiomes.

What's the relationship between breastfeeding and the microbiome?

Like all mammals, from elephants to orangutans, we have hair (or fur), we give birth to live young, and the females of our species produce milk to feed our infants. So far, so obvious. We are also all born in the same way: The babies all pass through the birth canal, which is loaded with specific bacteria, then they all drink their mother's milk.

Dr. Maria Gloria Dominguez-Bello describes the whole process of establishing the baby's microbiome, a process that starts in pregnancy, continues through birth, and is completed with breastfeeding: "The microbial communities that populate the mother's vagina at birth are lactic acid bacteria: bacteria very closely related to milk. The baby crosses the birth canal, which is full of bacteria, and then drinks milk for a long period of time. We think this is very important and is highly adaptive."

By "highly adaptive" she means that we have evolved over a long period of time so that this process happens exactly this way. This is how it's supposed to work. Breast milk contains all the nutrients the baby's body systems need to survive. But not just that: Breast milk also contains everything the gut microbes need to thrive. Survive and thrive; breastfeeding is the perfect natural system.

What's in breast milk?

A complex and complete meal, breast milk is specifically tailored to contain everything a baby needs over the first hours, days, weeks, and months of life. However, the first breast food is not milk at all; it is colostrum. A low-volume, thick, sticky, yellow-to-orange fluid, colostrum is highly concentrated, easy to digest, low in fat, and rich in protein.

Then, approximately three or four days after birth (or earlier if the woman has had children before), a breastfeeding mother will find that "her milk comes in." This is when she begins to "manufacture" mature breast milk. It's whiter in color, thinner, and produced in much greater volume than colostrum.

Both colostrum and breast milk are incredibly complex in terms of their components. They provide essential nutrients for the baby's growth and development including vitamins, minerals, fats, carbohydrates (sugar, principally in the form of lactose), amino acids, and proteins.[3] Both colostrum and breast milk also contain key immune components, including antigens, antibodies, and anti-inflammatories, as well as growth promoters to help stimulate cell growth, differentiation, and maturation.

As well as all that, both colostrum and breast milk contain a special type of carbohydrate that the baby can't digest—so what is it and why is it there?

We now know that this carbohydrate is a prebiotic oligosaccharide, and although the baby's gut can't break it down and digest it, the baby's gut microbes can. A prebiotic feeds the live, good bacteria (the probiotics) that live in the gut. Prebiotic oligosaccharides, delivered via breast milk, selectively stimulate the growth of the baby's good gut microbes so that they can multiply and mature. These, in turn, help to train the baby's immune system (see chapter 5).

How does the seed-and-feed process work?

In *Missing Microbes*, Dr. Martin Blaser says, "Once born, the baby instinctively reaches his mouth, now full of lactobacilli, towards his mother's nipple and begins to suck. The birth process introduces lactobacilli to the first milk that goes into the baby. This interaction could not be more perfect."[4]

Just like a precise recipe that produces beautiful food every time, it seems that for the optimal development of the baby's microbiome, you need the right ingredients (vaginal microbes and breast milk) mixed using the right method (vaginal birth and breastfeeding) for the perfect culinary result.

As a summary, here's the whole seed-and-feed process broken down step by step for vaginally born, breastfed babies:

1. During pregnancy, the baby develops in the womb in a near-sterile environment.
2. Late in pregnancy, the balance of the mother's vaginal microbiome shifts—there's a higher proportion of certain types of bacteria, such as lactobacilli, which are related to milk.
3. During vaginal birth, as soon as the waters break the baby becomes exposed to a sudden influx of vaginal microbes in the birth canal.
4. Vaginal microbes, including the high proportions of lactobacilli, coat the baby's skin and enter the eyes, ears, nose, vagina (if she's a girl), and mouth. The baby swallows some of the microbes, too.

5. Some of the bacteria arrive in the baby's gut.
6. The baby is born through the vagina. Immediately, he or she acquires more microbes, perhaps some from the mother's fecal matter, some from the air, and some from being touched—all these join the "colonization party" in the skin, oral, respiratory, and gut microbiomes. *This is the main seeding event for the baby's gut microbiome.*
7. The baby starts to breastfeed. The lactobacilli bacteria in the baby's mouth mix with breast milk and start to break down the lactose in the milk to produce energy for the baby. The baby swallows breast milk and, with it, some more microbes, including more lactic acid bacteria.
8. In the baby's gut, the bacteria are starting to grow hungry.
9. The food arrives! Breast milk provides nutrition and lactose (sugars) to feed the hungry baby, and special oligosaccharide (prebiotic sugars) that feed the hungry bacteria. *This is the feeding part of the process.*
10. With energy from the sugars, the lactobacilli and, in particular, the bifidobacteria (from the fecal matter and breast milk) multiply, quickly colonizing the baby's gut microbiome and preventing other harmful bacteria taking hold. These early arrivals in the gut also start training the infant immune system. A perfect process!

What happens to the microbiome in a formula-fed baby?

The subject of formula milk is both sensitive and emotive. We believe that everyone is entitled to make a personal choice about how they feed their baby—and the following is certainly not intended to make any parent feel guilty or worried about formula feeding. As parents who "topped up" their breastfed baby with formula milk, we know by personal experience that sometimes it can be very difficult to breastfeed, especially if a mother doesn't have the necessary level of one-to-one postnatal support. We know that some mothers desperately want to breastfeed but, like us, perhaps they may struggle to establish breastfeeding. Some mothers choose not to breastfeed at all, and others simply can't breastfeed.

Formula milk might meet the baby's nutritional needs, matching the nutritional content of breast milk, but some formula products may be missing other critical ingredients. Among those missing ingredients are hormones, antibodies, anti-inflammatory substances, specific strains of microbes, and the prebiotic oligosaccharides that feed the baby's gut microbes.

According to Dr. Neena Modi, "Formula these days is a very sophisticated product, but it can never, ever mimic all the constituents of breast milk. I think that mums can be assured that their babies get very good nutrition from formula feeding, but of course, what they won't get are the complex biological constituents that a baby receives from breast milk."

How does breast milk adapt to a baby's changing needs?

As the baby grows, so the composition of breast milk changes to accommodate the shifting physiological needs of the baby. This is true not only of the nutritional composition of breast milk but of its microbial content, too. In fact recent research by Raul Cabrera-Rubio of Teagasc, the Agriculture and Food Development Authority in Ireland, and colleagues indicates that the microbiome of breast milk changes quite significantly even within the first month.[5] Researchers found that certain species of bacteria were present in colostrum; then a month later, bacterial species typically found in the oral cavity greatly increased. This increase in specific bacterial species could be connected with the development of the infant immune system, as we'll explore later.

At the time of writing, the composition of formula milk doesn't reflect these complex microbial changes. A typical stage-one formula milk is described as being suitable from birth to six months.

This means a parent might give exactly the same formula to a newborn as they would give to a six-month-old baby. The baby's nutritional and microbial needs are not static over this time, so there is the potential that the unchanging formula milk doesn't deliver everything—including the microbes—that the baby needs for optimal health over this time.

Equally, there is the possibility that it is providing species of microbes too early in the development of the baby's gut health. Or, it might enable certain species to thrive at a time when those species might not be present at all if the baby were being breastfed, or not in quite the same proportions.

According to Dr. Dominguez-Bello, "We know that maternal milk reduces the diversity of the vaginal microbial communities that colonize the baby. So when you don't have maternal milk and you feed formula, what happens is the diversity increases a little. There are bacteria that shouldn't be there yet. They have arrived too early."

This difference between the microbiomes of breast- and formula-fed infants was also noted within the Canadian Healthy Infant Longitudinal Development (CHILD) birth cohort study. Professor Anita Kozyrskyj, principal investigator of the SyMBIOTA (Synergy in Microbiota Research) program, which derives its data from the CHILD Study, describes the microbial differences: "We found that, as others have, those infants that were exclusively breastfed, their species richness (this is a measure that we use to count the number of different bacteria) was lower than those infants that were exclusively formula fed. So formula-fed babies had the higher species richness. And the infants that were partially supplemented with formula—so some breast milk, some formula—their species richness was in between."

What this means is that, according to research, formula-fed babies, whether they are exclusively formula fed or given formula as a top-up to breastfeeding, have a more diverse microbiome at an earlier stage when compared with babies that are exclusively breastfed.

It's not just an issue of increased microbial diversity that comes too early. With breastfed babies, microbes in the baby's gut go through a maturing process. This might not happen in the same way with formula-fed babies. According to Dr. Rodney Dietert, "If a baby is not able to have breast milk for a long period of time as the main nutrient, then the microbes have not undergone the maturity process that occurs during infancy. And the adult microbial component has been shown to be quite different from those infants that have been formula fed versus those who have been breastfed. And that difference seems to be meaningful because those formula-fed children are at a greater risk for a variety of noncommunicable diseases."

In other words, the latest research indicates that a baby who has been exclusively formula fed from birth could be at increased risk of developing serious noncommunicable diseases (NCDs) later in life. Dr. Dominguez-Bello sums up this situation quite simply: "We don't know the magnitude of the consequences. If we break the rule of breastfeeding naturally, we may also be helping to compromise the education of the immune system of the baby." This potentially leaves the baby open to a greater possibility of disease.

In summary, the key points about formula milk and the microbiome are:

1. Formula milk provides nutrition for the baby.
2. Some types of formula milk may contain prebiotic oligosaccharides (special sugars) to feed the baby's microbes in the baby's gut, but other types of formula do not currently contain prebiotics.
3. Without the prebiotic special sugars, the colony of microbes in the baby's gut might not receive the specific food they need to multiply.
4. Formula milk doesn't currently provide the specific species of bacteria (found in breast milk) to perfectly match the needs of the developing baby over time.
5. Formula milk might be introducing some species of bacteria (that are not found in breast milk) too early. In effect, formula milk could be introducing bacteria that are not supposed

to be there, which could impact the optimal training of the baby's immune system.

6. If the formula-fed baby's immune system does not develop and mature as it would have if he or she had been breastfed, the baby could be at a higher risk of developing noncommunicable diseases later in life.

What are our personal reflections on formula versus breast milk?

With all our hearts, we wish we had known about all this research when it came to the birth of our daughter. When Toni struggled to breastfeed immediately after a C-section and received no support from the hospital staff, a nurse came in with a cup of formula milk, encouraging us to formula feed "for the sake of the baby." We gave in (it did feel like a sad defeat), and we gave our daughter some formula. Within just a few hours, the first food to reach our daughter's gut was formula milk, a decision that jars us still, years later. Then, over the next few months, Toni supplemented breastfeeding with formula feeding, which, according to emerging scientific evidence, could have had consequences for our daughter's developing immune system. It's too late for us to do anything about this now—we can't change the past. However, we feel that by writing this book, we are raising awareness of critical science that could help inform the parents of the future.

Since the time we formula fed our baby, a lot has happened in the world of formula milk. Products have become more specifically and nutritionally adapted to mimic breast milk, and scientists and manufacturers continue to make progress and improvements. Some formula manufacturers have already started adding prebiotics to their product range.[6] Perhaps in a few years, we'll see still further advances with personalized formula products tailored to the specific needs of every mother's unique microbiome, to better match all the components of breast milk.

Nonetheless, as we write, there is no perfect substitute for breast milk; no formula that delivers all the critical microscopic components the human cells and the microbes need for the optimal development of the baby (including his or her immune system) and of the baby's microbiome. According to Dr. Dietert, "Unless there are circumstances where the breast milk may be contaminated by environmental toxicants, prolonged breastfeeding is the best option. This involves the transfer of important immune components to the baby, but you also have the ideal food delivered for the microbial population."

What is the relationship between Cesarean section and breastfeeding?

Some researchers hypothesize that there is a correlation between difficulty in breastfeeding and having had a Cesarean section birth. Perhaps with the exception of elective Cesarean sections, this mode of delivery is often stressful for both mother and baby. Many women go into the hospital expecting to have a positive, natural birthing experience but, as a result of unforeseen circumstances, end up having emergency surgery to get the baby out. There may be lots of confusion, lots of strangers (more than in the standard delivery room), and a general state of high alert. An expectant mother is given an epidural or spinal anesthesia to paralyze the lower half of her body ready for the Cesarean section. For most women (or anyone) to suddenly lose almost all feeling from the chest downward is scary and unfamiliar. In exceptional circumstances, a woman having an emergency Cesarean section may need general anesthesia. But how might the operation affect her ability to breastfeed?

According to Professor Sue Carter, not feeling safe interferes with the natural biological processes of producing oxytocin, the hormone that is in part responsible for the "let-down" reflex that brings in the mother's milk supply. "The essence of the hormonal effects of oxytocin is probably a sense of safety. So when that system's working right, it helps you to feel safe, it probably helps the baby to feel safe. And in a state of safety, we can then interact with our offspring in just the right way. When we're afraid, we have trouble lactating."

Thinking about Toni's experience, we believe that immediate, one-to-one lactation support for someone who has just had a Cesarean section would be really helpful to initiate breastfeeding. (Indeed, that support should be available to all new mothers—regardless of the mode of birth.) Enough support to help the mother produce even a few drops of colostrum could make a big difference in terms of the seed-and-feed process. According to Dr. Modi, "Mums who deliver by Cesarean section do have a much more difficult time establishing their breastfeeding, but if they can produce even the few drops of colostrum and get that colostrum for their babies, they'll be setting their baby off on a fantastic start."

Here's a summary of the main points we've covered in this chapter:

1. The mother's breasts first produce a type of breast milk called colostrum; her actual breast milk comes in two to four days later.
2. Both colostrum and breast milk are very nutritionally complex, containing all the nutrients the baby needs to grow and thrive, including key components for the baby's developing immune system. Among them are antibodies, antigens, anti-inflammatories, and additional species of bacteria.
3. The bacteria that become dominant in the mother's vaginal microbiome late in pregnancy are lactobacilli, which are likely to arrive in the baby's gut first (from the baby receiving exposure to the microbial

payload in the birth canal) if he or she is born by vaginal birth. The lactobacilli help break down the lactose (sugars) in the breast milk, which gives the baby energy.

4. Both colostrum and breast milk contain carbohydrates (sugars) that are indigestible by the baby but digestible by the microbes that have newly arrived in the baby's gut (if the mother has had a vaginal birth). These are called prebiotic oligosaccharides (*prebiotic* meaning food to feed the good bacteria).

5. The oligosaccharides (sugars) give the bacteria energy to multiply and populate the gut. These are the ones that start training the baby's immune system.

6. Formula milk may contain all the nutritional requirements to feed the baby, but it may lack some of the prebiotics to feed the gut microbes. In addition, formula milk may not contain all the right bacterial species and immune components that the developing baby needs.

7. Because formula milk lacks some of the key microbe-related ingredients of breast milk, it could impact the training of the infant immune system, which could in turn result in health consequences later in the baby's life.

BREASTFEEDING AND THE MICROBIOME

— FOUR —

What Is the Impact of Cesarean Section on the Microbiome?

Until about one hundred years ago, a pregnant woman almost always gave birth at home. Now, in most industrialized nations, almost all women give birth in a hospital or midwife-led unit.

Undoubtedly, birth is far safer today than it was 100 years ago. In the United States in the early 1900s, between 600 and 900 women per 100,000 died each year of pregnancy-related complications.[1] By 1987, the statistics had fallen to 8 in 100,000 births.[2] (We should note, though, that the United States is one of only eight countries in the world where maternal mortality has actually increased in recent years. The latest statistics show that in 2013 maternal deaths in the United States had increased to 18 women dying per 100,000 births.)[3]

Professor Hannah Dahlen believes that it's not only the move into hospitals that has made birth safer. It's also, and perhaps more significantly, a result of improved hygiene, medical products, and general health. Among the influential factors when it comes to health is antibiotic medicine: "Birth came into hospital at the same time we had huge improvements in hygiene, in sewage, with women having fewer babies, with contraception, with blood products being available for transfusion, and with antibiotics being available. People immediately thought birth being in hospital was the reason for the better health outcomes, but it was all those associated factors."

What is a "typical" hospital birth experience?

As every mother, midwife, or doctor will tell you, no two births are the same. However, all over the world there are also plenty of defining similarities of hospital birth. If you group these similarities together, you start to paint a picture of what modern birth can look like for many women.

Doctors, midwives, and other healthcare professionals measure a woman's labor in terms of its "progression." The speed of the baby's descent through the birth canal, the speed of cervical dilation, and the frequency and length of contractions are all measures of how labor is medically deemed to be progressing.

In essence, then, hospital staff continuously assess a laboring woman in terms of the speed of the various labor markers. If a woman's body is thought not to be working fast enough (according to the parameters set out by the hospital and good-practice guidelines, and the medical team's judgment), doctors may make the decision to intervene to speed along the baby's birth.

There are several ways in which this can happen. One way is for a doctor or midwife to artificially rupture the membranes (also known as breaking the waters) to get things going. As we explored in chapter 2, this immediately exposes the baby to an influx of microbes that begins the main seeding of the baby's microbiome.

If labor still doesn't quicken up, the mother may be given an intravenous drip of synthetic oxytocin. According to Professor Dahlen, as a result of the synthetic oxytocin drip, "contractions are more painful; you may not be releasing the same amount of endorphins as you might have done without these interventions."

With contractions becoming more painful, midwives may suggest or the mother may request an epidural to help deal with the pain. Professor Dahlen continues, "You have an epidural, but you can't move around. The baby is no longer getting the assistance of your movements to help it navigate the birth canal. The heart rate will often plunge. That means we either have to rush you to theater or we do a forceps delivery, and so the cascade of intervention goes on."

In modern maternity-care practice, most women have at least one, if not multiple interventions. According to Professor Dahlen, "Major interventions like induction, speeding up of the labor, giving you an epidural, giving you antibiotics, forceps, Cesarean; the way we view these in society today is almost as though they are more normal than normal birth. They are the tools of our trade. They are us doing something." To give an idea of how common these interventions are, births in England in the year 2013–14 show that 26.2 percent were by Cesarean section (13.2 percent were elective, 13 percent were emergency), 25 percent were induced, and 12.9 percent were instrumental (delivery by forceps or ventouse, which is a suction device applied to the baby's head to assist the birth).[4]

To analyze the potential impact of all possible labor interventions in relation to seeding the baby's microbiome would be beyond the scope of this chapter. However, we will look at two very common labor interventions: the use of synthetic oxytocin and of Cesarean section.

What is the impact of synthetic oxytocin on the baby's microbiome?

Professor Sue Carter, one of the world's foremost experts on oxytocin, provides a useful historical perspective on the use of oxytocin's synthetic cousin, called Pitocin in the United States; Syntocinon in the UK: "The original idea of Pitocin was that it was going to 'cause' birth. But it quickly became obvious that it wasn't sufficient to create a birth. What was actually happening was that the uterus was constricted and the baby's head, if it was pointed down as it should be, was hitting against the cervix. Eventually this would lead to a cascade of other hormones, and with that cascade, birth would in theory occur. But for many women given Pitocin over the years, their cervix and their pelvis were not prepared. They really weren't primed by the other hormones that set this up. So instead of getting a birth, their babies were stuck. This led to the need for some kind of procedure like forceps, episiotomy, or Cesarean section."

If used correctly, synthetic oxytocin can help stimulate a woman's labor. If used after the birth, it can also help reduce bleeding. As it can be very effective in these aims, synthetic oxytocin is now very commonly used in hospitals all around the world.

To be very effective, though, the synthetic oxytocin needs to be administered in the right proportions at the right time. According to Professor Carter, "I think if it's used properly, synthetic oxytocin may be able to stimulate the process. But if it's given too fast and too much in the beginning, the natural system is not brought into it and labor may actually stop."

If labor stops because of an inappropriate dose of synthetic oxytocin, or for any other reason, the birthing mother may need a Cesarean section in order for the baby to be born safely. As we'll see later in this chapter, this Cesarean section could interfere with the optimal seeding of the baby's microbiome.

Lack of high-quality, large-scale research means that we don't yet know whether or not administering synthetic oxytocin during labor and birth can affect the baby's long-term health in other ways. With a high percentage of women in hospitals being given Pitocin or Syntocinon at some point in their labor and birth, there are very few women left to study who haven't had the drugs.

According to Professor Carter, there may be other reasons too. "Historically, birth interventions were rare, but now they're very common. It's very expensive to do a longitudinal study in people because we're a slow-developing species. We have to wait 20 years before we're more or less fully grown. There's a third problem, I'm afraid. I'm not sure we want to know about the bad things that might be coming from the birth interventions we're doing." Professor Carter makes a very important point. Studying the potential long-term effects of the use of synthetic oxytocin in childbirth might be costly, and it might take time, but it might also reveal uncomfortable truths about the potential long-term damage that has been caused to children whose mothers were administered the drug even many decades ago.

How has the rate of Cesarean sections changed over time?

Cesarean section can be a lifesaving surgical procedure. Many thousands of women and babies, perhaps millions, wouldn't be alive today if it weren't for the baby being delivered surgically. However, rates of Cesarean section have increased to such an extent that surgical birth has now become relatively common, despite the fact that in 2015 the World Health Organization (WHO) released a statement saying that "Cesarean sections should be performed only when medically necessary."[5]

The rapid rise of Cesarean section as a way to give birth raises questions for many scientists, including Dr. Martin Blaser: "In the USA, 20 years ago, the C-section rate was 18 percent, now it's 32 percent. So the question is, what drove that big increase? Do all those women need it? In some communities in Sweden, the C-section rate is only 5 percent. But in other parts of the world, it's 50 percent or higher."

At the time of writing, the latest statistics from the from the Organization for Economic Cooperation and Development (OECD) state:[6]

- In the UK, around one-quarter of all births are by Cesarean section

- In the United States and Australia, around one-third of all births are by Cesarean section
- In Turkey, more than half (50.4 percent) of all births are by Cesarean section

In Brazilian private hospitals, the rate is even higher. According to a 2015 statement by the Brazilian Minister of Health, Arthur Chioro, 84 percent of the country's births in private hospitals are by Cesarean section.[7] Chioro said in a statement, "The epidemic of Cesareans we see today in this country is unacceptable and there is no other way to treat it than as a public health problem."

Of course, such "epidemic" rates are not true of other parts of the world. In some sub-Saharan African countries, for example, there is a lack of access to Cesarean sections. Women who desperately need the procedure to save their own or their baby's life aren't able to have it. This may be a contributing factor to high rates of mothers and their babies dying in childbirth in some developing nations.

Why have Cesarean sections become so prevalent?

Advances in medical technology and medication—anesthesia, surgical techniques and processes, blood transfusions, antibiotics, and suture materials—mean that Cesarean sections are now safer than ever, improving patient outcomes.

The Misgav Ladach Cesarean section technique, a type of straight transverse surgical incision, from side to side, just below the bikini line, has enabled (according to a 1999 academic paper) "quicker recovery, less use of post-operative antibiotics, antifebrile medicines, and analgesics. There is a shorter anesthetic and shorter working time for the operative team. It is suitable for both emergency and planned operations."[8]

Cesarean sections are now generally perceived as a relatively safe alternative to vaginal birth. As suggested by Professor Philip Steer, "My personal view of why the C-section rate has been rising steadily ever since I qualified is that we keep making Cesarean sections safer. And because they're safer that means the threshold of risk in labor at which we would resort to a Cesarean section gets lower and lower and lower."

Many of the Cesarean sections performed in industrialized nations occur before the start of labor. The latest statistics available for England show that more than half of all Cesarean sections were elective (see page 73). Some of these occur because the mother or baby has complex health problems; others occur because the mother has chosen to have her baby delivered surgically.

Many women may have compelling reasons to choose a surgical rather than a vaginal birth, including a fear of or anxiety about vaginal birth. The UK's maternity system supports the notion that a woman can choose to have a Cesarean section birth. In the UK, under National Institute of Health and Care Excellence (NICE) guidelines that were new in 2011, "For women requesting a C-section, if after discussion and offer of support (including perinatal mental health support for women with anxiety about childbirth), a vaginal birth is still not an acceptable option, offer a planned C-section."[9]

Once a woman has had one Cesarean section, a doctor may recommend another surgical birth for a subsequent baby based on a calculation of risk relating to complications. A vaginal birth after Cesarean section (VBAC) might be medically possible, but it might not always be a fully supported or viable option at some hospitals. A rise in expectant mothers deemed "high-risk," as well as an environment in which a health professional could fear litigation if he or she does not perform a Cesarean section quickly enough, might contribute to the increase in the numbers of Cesarean section births.[10]

What is the relationship between a Cesarean section and long-term health?

Statistically, Cesarean section might be a relatively safe procedure for mother and baby, but clinicians such as Professor Steer are starting to become more aware of the potential long-term problems associated with surgical birth: "Until recently, the main emphasis with Cesarean section has been to avoid the damage of difficult labor, because the baby can get short of oxygen during a difficult labor. The whole objective of Cesarean sections is to deliver babies in good condition, and obstetricians have generally thought 'Well, that's my job over' when they pass the baby onto the neonatologist. However it's become clear that there are some disadvantages to the baby of not having been born by a normal vaginal birth."

Some of the disadvantages of surgical birth relate to alterations in the seeding of the baby's gut microbiome. In order to illustrate how the baby could miss out on vital microbial exposure, it might be useful to describe what happens during conventional Cesarean section.

What is the Cesarean section procedure?

Typically, the mother lies on an operating table, with a curtain providing a shield between her and the area of incision. Quite often this curtain is made from the mother's own surgical gown. She is usually given an epidural, a drip into her spinal cord that numbs the lower part of the body. If there are certain indications, she may need to have a general anesthetic.

Doctors make a surgical incision to slice open the lower abdomen and cut through the various layers of the abdominal wall, to reveal the lower portion of the uterus (the womb). They then make a further incision in the uterine wall. Usually this cut is horizontal (from side to side), but in certain circumstances it may be vertical (up and down). For an elective Cesarean section, once the incision has been made, the baby is revealed enveloped in the amniotic sac. For an emergency Cesarean section (which occurs only when there's a problem during active labor), if the waters have already broken, the baby may be revealed within a split amniotic sac.

The amniotic sac is then split open (if necessary), and the obstetrician pulls the baby from the mother's womb either using their hands, or a pair of forceps. (According to Toni, this sensation felt like a whole load of wet washing being yanked out of a top-loading washing machine.) The umbilical cord is clamped and cut and the baby is independent from his or her mother. At this point it's common practice for the baby to be quickly shown to the mother before a neonatologist or nurse takes the baby off for evaluation. This might mean that the baby is taken away to a resuscitation table on the other side of the operating theater. Then the baby might be cleaned, tagged, weighed, and swaddled in a towel.

Meanwhile, the surgeon detaches and removes the placenta from the mother's uterine wall. The uterine incision is closed using stitches or staples, followed by closing up of the abdomen.

The "birth" part of the procedure happens relatively quickly: From the first incision to the baby emerging into the outside world usually takes around five minutes. It might take another forty minutes to an hour for the surgeon to repair the uterine and abdominal walls.

While she's being stitched up, the mother usually gets to hold her baby for the first time—often a full ten minutes after the baby has been born. In some places around the world, the baby might then be taken to a nursery while the mother rests in a separate recovery room.

What about seeding a Cesarean section baby's microbiome?

Because he or she is pulled directly from the mother's abdomen, a baby born by Cesarean section does not pass all the way through the birth canal and so is likely not to have received the full complement of lactobacilli–rich vaginal microbiota acquired by babies who've been born vaginally.

Dr. Maria Gloria Dominguez-Bello, a microbiologist engaged in pioneering research into how mode of birth impacts the gut microbiome, says, "People believe that babies born by C-section are exactly the same as babies born vaginally. But now with molecular methods of sequencing, we can see that babies born by C-section lack the vaginal microbiota that vaginally born babies have as they did not pass through the birth canal."

But it's not just that babies born by Cesarean section may lack vaginal microbes. A baby born by Cesarean section is likely not to have come into contact with the mother's fecal matter during the birth process, so the baby wouldn't be exposed to the mother's intestinal microbes. In effect, a baby born by Cesarean section is likely to miss out on receiving the special payload of the mother's vaginal and intestinal microbes. These microbes are supposed to be the first arrivals at the gut microbiome "colonization party." As we'll see in the next chapter, a lack of exposure to them could impact the optimal training of the infant immune system.

Up until the point that the amniotic sac is split open, the baby is in a near-sterile environment. As soon as the surgeon splits open the sac, a deluge of microbes floods over the baby. In a Cesarean section birth, those critical first microbes come from the air of the operating theater, not the birth canal. As identified by Dr. Dominguez-Bello: "When we examine babies within seconds after being birthed, we observe that they have microbes that they get from the air of the surgery room."

That's when the main seeding event begins for a Cesarean section baby. As soon as the amniotic sac is opened, airborne microbes start colonizing the newborn's skin, entering the baby's eyes, ears, nose, and mouth.

During her research, Dr. Dominguez-Bello found that these airborne microbes come from the skin of someone, or more than one person, in the operating theater. According to Dr. Dominguez-Bello, "These skin microbes are not from the mother; probably the mother is the least important source. They are from another person, or other people, in the theater."

With each physical contact the baby acquires more microbes—as a result of being handled, being treated with medical instruments and implements, being wiped with a towel, and having skin-to-skin contact with the mother or father. Of course, the same is true once a baby has been born vaginally—the crucial difference is that, for a baby born by Cesarean section, the mother's vaginal and intestinal microbes aren't present at the gut microbiome "colonization party" alongside the microbes from the environment.

Dr. Dominguez-Bello sums up the seeding of the Cesarean section baby's microbiome: "Babies born by C-section do not acquire vaginal microbes at birth because they don't cross the birth canal. But they do get a set of bacteria that corresponds to human skin, which they get from the air of the operating theater."

Is there a difference in microbial exposure between elective and emergency Cesarean sections?

The simple answer to this question is, probably, yes. However, there's limited research on the exact difference in microbial exposure between Cesarean section birth with and without labor. Whether or not the baby is exposed to the mother's vaginal microbes during the birth might depend upon whether the mother's waters have broken and if the baby enters the birth canal prior to the Cesarean section surgery.

If the waters break early, and if the woman is fully dilated and has been in the second pushing stage for some time (with the baby in birth canal) before needing an emergency Cesarean section, the baby might have received some exposure to the mother's vaginal microbes while in the birth canal, even if the mode of birth then changes.

Here are two examples to illustrate how microbial exposure might differ for babies who are born by Cesarean section with or without labor.

First, Cesarean section with labor. An expectant mother's waters break sometime during active labor, and her labor progresses until her cervix is fully dilated. The mother feels the urge to push, and the baby moves through the birth canal. It's the mother's first baby, and the second stage is taking some time, maybe upward of an hour. The mother tries different positions to keep the labor progressing. The attending obstetrician becomes concerned that the baby is in distress and suggests an emergency Cesarean section, which goes ahead. In this example it's perfectly reasonable to suppose that the baby has received some exposure to the mother's vaginal microbes in the birth canal before he or she is born by Cesarean section.

Second, an elective Cesarean section (also known as Cesarean section without medical indication or Cesarean section by maternal request), which is an example of Cesarean section without labor. This usually happens before labor begins, perhaps even before the baby's due date. The waters have not broken by the time the woman goes into surgery, and the amniotic sac is still intact. In this case we might suppose that the baby, who has remained in the near-sterile environment of the sac until the moment of birth itself, does not receive any exposure to the mother's vaginal (or intestinal) microbes during birth.

There is a third scenario, of course—a Cesarean section with labor, but when the woman goes into surgery before her waters have broken. We have limited research to draw from for this, but we might suppose that if the baby has not entered the birth canal and the waters have not broken, the baby might not receive any exposure to the mother's vaginal (or intestinal) microbes before the Cesarean section surgery occurs.

We need more research, but at the time of writing, scientific investigation indicates that all babies born by Cesarean section have an altered microbiome compared with babies born vaginally. In any Cesarean section situation, the baby is likely not to have received the mother's vaginal and intestinal microbes at the levels we might expect from vaginal birth. Not just that, but the fact that the first microbes to arrive in and colonize the baby's gut microbiome are likely not to originate from the mother's vaginal or fecal microbiome could impact the training of the infant immune system, with consequences for the baby's long-term health.

What are the differences between microbial species on a Cesarean section baby compared with a vaginally born baby?

Much of the published research focuses on comparing the microbiome of a vaginally born infant to the microbiome of a Cesarean section–born infant in broad terms. In essence, birth is usually divided into two categories:

VAGINAL BIRTH   CESAREAN SECTION BIRTH

Sometimes as well as mode of birth, research also factors in how a baby is fed—so whether the baby was breast- or formula fed. If we tabulate the options, we have four possible scenarios:

VAGINAL BIRTH—	CESAREAN SECTION BIRTH—
BREASTFED	BREASTFED
VAGINAL BIRTH—	CESAREAN SECTION BIRTH—
FORMULA FED	FORMULA FED

In most of the current research available, there's little mention of the exact circumstances surrounding the birth. For example, in a vaginal birth we often don't know if the baby was born at home, at a birth center, or in a hospital; when (and if) the mother's waters broke; how long the mother was in labor before her waters broke, if they did; whether the mother had a water birth, and if she did, when she got into and out of the water; whether she had any antibiotics immediately before birth (for example, if she had tested positive for group-B streptococcus). In Cesarean section birth we rarely know whether the surgery was elective or emergency;

whether the mother had been given antibiotics before her surgery; when and if the waters had broken; whether or not the woman was in active labor before surgery; and what medical interventions she had received before going into labor. All these factors, in both vaginal and Cesarean section births, could influence the seeding of the baby's microbiome.

In the future, research will focus on more pieces of the birthing puzzle, and we will be able to build up a more nuanced picture of exactly how the mode of birth impacts the microbiome of the infant, but at the moment, we have only limited research resources to give us answers. Dr. Dominguez-Bello explains that when we analyze the microbiome of newborns, "Babies born vaginally have mostly bacteria such as lactobacillus and also bifidobacterium," which are types of anaerobic bacteria (they can live without air).[11] Cesarean section babies, on the other hand, have a microbiome that contains more "streptococcae and staphylococcus. These are mostly aerobic bacteria (so they live in the presence of air), typically from the human skin, very different communities than the vaginal bacteria." Here's our interpretation of Dr. Dominguez-Bello's pioneering research illustrated in a very simple two-square table.

VAGINAL BIRTH	CESAREAN SECTION BIRTH
More lactobacillus and bifidobacterium	More streptococcae and also staphylococcus

In the CHILD Study, the cross-Canadian team of researchers, including Professor Anita Kozyrskyj, also found that there was a difference in the bacterial footprint depending on mode of birth.[12] According to Professor Kozyrskyj, "In our pilot study, when we compared the infants that were delivered by Cesarean compared with those delivered vaginally at three or four months of age, those children that were delivered by Cesarean, they definitely had lower percent composition with this genus *Bacteroides*. At 1 year of age, it was lower but not as dramatically so."

84

What is the impact of antibiotics on a baby's microbiome?

The situation regarding potential alterations in the gut microbiome is further complicated if the expectant mother receives antibiotics during pregnancy, labor, or birth. A common reason for antibiotic administration is to combat the baby's risk of infection by group-B streptococcus, otherwise known as group-B strep or GBS.

What is group-B strep?

Group-B strep is a commonly occurring bacteria. In the UK, it is present in the microbiome of between 20 and 30 percent of adults (both men and women). And 22 percent of pregnant women carry group-B strep in their vaginal microbiome.[13]

Although most babies exposed to group-B strep during vaginal birth will suffer no side effects, there is still a significant risk that a newborn baby could develop a group-B strep infection. A serious medical condition, group-B strep infection can lead to meningitis, pneumonia, and sepsis (blood toxicity). In the UK, about 1 in 2,000 babies develops a group-B strep infection shortly after being born. Around 1 in 10 of those babies will die.[14]

The UK health service takes a "risk-based approach" to group-B strep—meaning that expectant mothers are not routinely tested for the bacteria, although the infection might be detected if a pregnant woman provides urine or vaginal samples for other reasons. There are certain risk factors that midwives and doctors will look for to pick out women at risk. Having had a previous baby with group-B strep infection, going into premature labor, having a temperature higher than 37.8°C (100°F) during labor, and breaking waters more than 18 hours before the birth are all common indicators.[15] If a mother has one or more of these risk factors, her doctor could prescribe antibiotics to be taken either when the waters break or from the start of labor, whichever comes first.

Doctors in the United States, on the other hand, take a "universal approach." This means that all pregnant women receive screening for group-B strep at 35 to 37 weeks, and those who test positive take antibiotics during their labors.[16]

What are the consequences of receiving antibiotics during Cesarean section?

Every surgery carries with it a risk of infection. It's common practice for doctors to give mothers a shot of antibiotics before a Cesarean section in order to reduce the risk of infection as a result of the surgery. Dr. Dominguez-Bello covers this in her research into the effects of Cesarean section on the microbiome: "C-sections also involve antibiotics. The mother gets one gram of penicillin before birth so her microbiota is impacted, and during the birth, the baby will be receiving those microbes that have been impacted by antibiotics."

What do we learn by analyzing the effects?

Professor Kozyrskyj and her colleagues have recently published new research on maternal antibiotics and their impact on the infant gut microbiome as part of the CHILD Study (see "What are the differences between microbial species on a Cesarean section baby compared with a vaginally born baby?" on page 83).[17] This research investigated the effects on two hundred Canadian infants born between 2010 and 2012. Researchers found that 21 percent of expectant mothers received antibiotics because they were at higher risk of group-B strep or because of pre-labor rupture of membranes. A further 23 percent of mothers received antibiotics because they had either an elective or an emergency Cesarean section. That means that, between 2010 and 2012, a considerable proportion of Canadian expectant mothers received antibiotics during the intrapartum period, the time around labor and birth.

The researchers found that the infant gut microbiome changed in relation to the intrapartum antibiotics. All infants, whether born vaginally or surgically, had a significantly altered microbiome at age three months, when compared with babies whose mothers had not taken antibiotics. In terms of specific bacteria, at age three months, the infant microbiome of antibiotic-exposed babies had fewer bacteroides and parabacteroides (which are important for numerous metabolic activities and may help protect against harmful pathogens)[18] and more enterococcus (lactic acid bacteria recognized as a human pathogen with a high level of intrinsic antibiotic resistance)[19] and clostridium. The researchers found that the differences persisted for up to twelve months in Cesarean section babies, particularly in babies who were not breastfed.

Professor Kozyrskyj and her colleagues conclude that "intrapartum antibiotics in Cesarean and vaginal delivery are associated with infant gut microbiota dysbiosis, and breastfeeding modifies some of these effects. Further research is warranted to explore the health consequences of these associations."

A "dysbiosis" in the infant gut microbiota associated with maternal antibiotics means that the baby's gut microbiome is not quite "in balance." In other words particular bacterial species that would have been there if antibiotics had not been taken are missing or not present in optimal quantities—there may be too many of one species and too few of another.

Note for expectant women: Always discuss with your doctor the benefits and risks of taking antibiotics during pregnancy, labor, and birth.

Why does an altered infant gut microbiome matter?

As the differences between the respective microbiomes (between vaginally versus Cesarean section–born babies, or maternal antibiotics versus no antibiotics, or breastfed versus formula-fed babies) become less apparent as infants grow older, a skeptical observer might see similarities in their respective microbiota profiles as adults and think, "Well, their bacterial profiles look pretty similar now, so what's the problem?"

It's a good question. What exactly is the problem with an altered infant gut microbiome in the first twelve months if those differences don't persist? Why does it matter?

Professor Kozyrskyj has hypothesized that the infant's microbial footprint changes over time: The baby's bacterial profile at birth is different from their bacterial profile at a few weeks, which is different from their profile at a few months, which is different from the profile at a year. She goes on to suggest that the infant gut microbiome develops in a special "successive" pattern: "The development of the infant gut microbiota occurs successively. So you have the founding bacteria, and they lay the foundation for the next bacteria to colonize. The next bacteria that colonize are the ones that we have as an adult. So the genus *Bacteroides*, this is one of those bacteria that is next to colonize."

It's not only the abundance or lack of abundance of a particular species at one particular time in the infant's early life that is significant. It's the pattern of colonization over the course of the baby's first year. According to Professor Kozyrskyj, "I don't think it's as simple as saying, 'a footprint at three months is important,' or 'a footprint at 1 year is important.' In our opinion, my opinion and my colleagues', it's the pattern over the first year of life."

This seems to suggest that certain bacteria are supposed to arrive in a certain order at a certain time as part of a complete colonization pattern that happens over the first year of life. By "supposed to," we mean that this is the ideal order based on research. We know that the first bacteria to arrive in the gut microbiome are believed to be critical for optimal training of the infant immune system.

Cesarean section, maternal antibiotics, and even formula feeding could disrupt this whole pattern of successive colonization right from the moment of the baby's birth. The next question, therefore, has to be: Could this have consequences for a child's health later in life?

Professor Kozyrskyj and her colleagues are currently exploring whether or not there is a potential link between interference in successive microbial patterns over a baby's first twelve months and the development of certain health conditions, including allergies. They recently published a paper looking at the connection between the colonization patterns in the infant gut microbiome and food sensitization over the first year of life.[20] The CHILD Study investigators concluded, "Low gut microbiota richness and an elevated enterobacteriaceae/bacteroidaceae ratio in early infancy are associated with subsequent food sensitization, suggesting that early gut colonization may contribute to the development of atopic disease, including food allergy."

In answer to the question about whether or not interference in microbial patterns does have health consequences in later life, this paper suggests that it could. If certain specific bacteria are supposed to arrive in the infant gut at certain specific times in a certain specific order as part of a complete colonization pattern, any interference in this pattern could well be a contributing factor to a child being more susceptible to developing food allergies.

What can you do if you need a Cesarean section?

When we learned that a baby born by Cesarean section may lack species of microbes that would have been acquired if they had been born vaginally and that this could have lifelong consequences for the immune system, we immediately thought back to the circumstances of our daughter's C-section birth.

After four failed inductions, Toni had a C-section. With each induction attempt, Toni started having contractions but not fully established labor. With each attempt, the cervix started to open, only for the labor to stall. The waters did not break at any point, which would suggest that our child probably had no exposure to Toni's vaginal microbes.

We've never had our daughter's gut microbiome sequenced, so we don't know whether her microbial footprint still reflects the circumstances of her birth.

Knowing what we know now, we would have included a list of things in our birth plan to make provisions for microbial exposure in the event we needed to have a C-section. This list would have included asking the medical staff to lower the screen at the moment of birth, for the baby to be delivered directly onto Toni's chest for immediate skin-to-skin contact in the operating theater, with exclusive breastfeeding from that point on. If Toni's vaginal microbiome was screened during pregnancy and tested negative for pathogens, we might also have considered "swab-seeding." This procedure for elective Cesarean sections (with strict protocols) is currently being researched by Dr. Dominguez-Bello.

Much of this list would have been possible thanks to a new type of "woman-centered" Cesarean section now available in some progressive hospitals in the UK and in some other parts of the world, including in the United States.[21]

Sometimes called natural Cesarean section or gentle Cesarean section, the techniques involve small but significant changes to the traditional Cesarean section procedure, including small changes to working practices, additional staff (for example, an extra nurse or member of the neonatal team), and positioning of the intravenous drip in the mother's nondominant hand so that it is easier for her to hold the baby.

What are the benefits of skin-to-skin contact after a Cesarean section?

According to Professor Steer, "In enlightened obstetric practice, it's very normal to deliver the baby out of the tummy and give the baby straight to the mum so you can have skin-to-skin contact."

As well as facilitating mother–baby bonding, skin-to-skin contact enables the critical transfer of skin microbes between mother and baby and helps to establish breastfeeding. Provided there is not a medical emergency for either mother or baby, the mother might not need to be separated from her baby at all. Medical staff can leave the umbilical cord attached to the baby, allowing it to stop pulsating before it is cut, all the time the baby is on the mother's chest. Of course, this is the same for babies born by Cesarean section as for babies born vaginally.

If a mother can't have immediate skin-to-skin contact because of a medical emergency (whether she's had a vaginal or Cesarean section delivery), the baby could have skin-to-skin with the father or other close family member. We have little research to draw from with regard to microbial transfer in this scenario, but we can suppose that it would be beneficial. Microbes from the biological father or close family member are those literally related to the baby, and they represent the microbes that the baby will have all around them when the family gets home.

There could be other emotional and psychological benefits of skin-to-skin contact between the newborn and someone other than the mother, as described in a moving story told by Lesley Page. "I remember a really lovely young couple. They were from India, living in the UK. And the mother had an emergency Cesarean section. After the C-section, she was just stunned. I looked at the father who didn't quite know what to do, and I said to him, 'Can we put the baby next to your skin?' And he lifted up his scrubs and put the baby next to his chest, and I said to put the baby over his heart. And his face lit up. And I felt that must have made a difference to that father and baby relationship in the future."

What are the benefits of breastfeeding after a Cesarean section?

In chapter 3, we learned that breastfeeding is a delivery system for nutrition, antibodies, anti-inflammatory substances, immune components, hormones, prebiotics, and specific strains of bacteria to the baby. These could all be of particular benefit for all babies born surgically.

In the CHILD Study paper on the impact of maternal antibiotics, the researchers make the point that breastfeeding modifies some of the dysbiosis effects associated with the mother taking antibiotics during labor and birth.[22] "Thus, our results demonstrate the benefit of continued breastfeeding after emergency C-section in promoting a post-weaning gut microbiota profile comparable to vaginally born infants without IAP exposure." (*IAP* stands for "impact of maternal intrapartum antibiotic prophylaxis," referring to the taking of antibiotics in the period surrounding labor and birth to prevent infection.) Their research clearly shows the benefits of breastfeeding for the infant microbiome, particularly for babies that have been born by emergency Cesarean section.

According to Professor Carter, "One of the body's protective mechanisms is to breastfeed the baby. Lactation, in my mind, is a kind of insurance policy that can follow any kind of birth. It's more difficult if there's a C-section. So by getting the C-section rate up so high, we're making it harder for women to lactate. And that's unfortunate because lactation is your extra insurance for that newborn baby."

If all babies, mothers, and fathers were routinely supported to have immediate skin-to-skin contact with extra support for breastfeeding, and if this were the case for all vaginal and Cesarean section babies, we would probably see an increase in breastfeeding rates, which would confer microbial benefits to the baby. In turn, this could potentially significantly improve the baby's lifelong health.

What is swab-seeding?

Dr. Dominguez-Bello is researching a pioneering swab-seeding technique that might have appeared on Toni's birth plan (if it was available back then), especially if Toni's vaginal microbiome tested negative for pathogens during pregnancy and if Toni fit the inclusion criteria and strict protocols of Dr. Dominguez-Bello's research. The technique is an answer to a problem described by Professor Steer: "I lecture quite a lot about Cesarean section, and I'm frequently bringing up the topic of the microbiome and its importance for the future development of the baby. I think it's likely that we will find techniques, for example, by getting a selection of the mother's bacteria and giving them artificially to the baby, making sure that the natural process is mimicked."

The swab-seeding technique aims to artificially seed a Cesarean section baby's microbiome with the mother's vaginal microbes. In the technique, a clinician wipes the baby's face with a swab soaked in microbes taken from the mother's vagina. According to Dr. Dominguez-Bello, "Considering that there has to be a C-section, we thought, what can we do to restore these babies? And it sounds very logical. The idea is if the baby needs to be born by C-section, we deliver them by C-section and then show them the vaginal contents of the mother."

To date, Dr. Dominguez-Bello has conducted her research in Puerto Rico, Ecuador, Bolivia, and Chile. Recently, permission has been granted for her to carry out further investigations in other countries, including Sweden and the United States, with other teams in other countries lining up to be part of the research.

However, we're still at an early stage in monitoring the effects of swab-seeding, and presently the cohort group includes only babies born by elective Cesarean section and whose mothers have been carefully screened to make sure their vagina is healthy prior to the procedure.

How does swab-seeding work?

To begin swab-seeding, researchers insert a sterile tampon-like swab into the mother's vagina immediately before surgery, then remove the swab and put it into a sterile dish while the Cesarean section itself is under way. As soon as the baby is born, researchers wipe the swab, first over the baby's mouth, then over the rest of their face and body.

Dr. Dominguez-Bello is now analyzing the first set of results: "We will publish the first month of data to show the principle works. We are following these babies for 1 year, and we are still recruiting. So far, with the babies we have inoculated, we have had no health issues at all. In 2 or 3 years, we will produce a bigger paper showing more, including data about the health of the baby."

We should exercise a strong note of caution given that it's very early days in the life of this research project, but the preliminary results look favorable. To quote from an article about Dr. Dominguez-Bello's research published in *Nature* February 1, 2016: "The four babies who received the swabs harbored skin, gut, anal, and oral bacterial communities that were more like those of infants delivered naturally, compared to the C-section-delivered babies who did not go through the procedure."[23] This means that the swab-seeding technique may partially restore the microbiome of Cesarean section babies, enabling the microbial profile of Cesarean section babies to be closer to that of vaginally born babies. Dr. Dominguez-Bello refers to those babies that have received the swab-seeding as having been "inoculated."

She says, "We have only analyzed at the level of the communities, but so far, we can show that inoculated babies have microbial clusters closer to vaginally born babies. In other words we do have preliminary evidence that we are at least partially restoring the vaginal microbes to C-section delivered babies. Although they were born by C-section, they look more similar to vaginally delivered babies than to C-section delivered babies."

Dr. Dominguez-Bello's research is ongoing, and at the time of writing, it's not known whether or not the swab-seeding technique does actually confer positive health benefits for the baby. But if the full results confirm that swab-seeding can at least partially restore the microbiome of babies born surgically, and if it's shown that this could have positive health consequences for the baby, this technique could be a game-changer, at least for those mothers that fit the strict inclusion criteria of Dr. Dominguez-Bello's research. The swab-seeding technique may never replace the full benefits of having a vaginal birth, and indeed, the technique does not replace or restore the mother's intestinal microbes.

Perhaps, one day in the future, the swab-seeding technique could become a partial restorative option for all babies born by Cesarean section the world over. Until that time, it's worth emphasizing that swab-seeding is not currently a recommended standard medical procedure—it's simply a procedure being scientifically investigated right now using strict protocols. If a baby is swab-seeded without these strict protocols being observed, there is a risk that harmful pathogens harbored in the vaginal microbiome may be introduced to the baby with unintended consequences (particularly if the mother's vaginal microbiome has not been screened during pregnancy). Hopefully before too long, we'll have more definitive answers about this technique, which will no doubt prove very helpful to expectant mothers and their healthcare providers.

Here's a summary of the main points we've covered in this chapter:

1. Over the past 100 years, birth has become more medicalized. Common interventions include induction, artificial rupture of membranes, use of synthetic oxytocin, epidurals, and Cesarean sections.
2. To date there have been few studies into the possible long-term impact on the health of the babies born using these interventions in childbirth.
3. With an elective Cesarean section (or a Cesarean section performed before the waters have broken), the baby is likely not to be exposed to the mother's vaginal and intestinal microbes—exposures that the baby is likely to have received if born vaginally.
4. With elective Cesarean section, the baby's first main exposure to microbes will come from the air of the operating theater, probably from the skin of someone in the operating theater, possibly not from the mother at all.
5. With the type of emergency Cesarean section where the mother is in active labor and her waters have

broken prior to surgery, the baby could have received some exposure to the mother's vaginal microbes in the birth canal.

6. As with a baby born vaginally, a baby born by Cesarean section (elective or emergency) would then acquire more microbes from being touched and fed—this influx of microbes would join the "colonization party" in the baby's gut.

7. Neither babies born by elective nor emergency Cesarean section are exposed to the full complement of the mother's vaginal or intestinal microbes, which means the infant is likely to have an altered gut microbiome when compared with babies born vaginally. This could impact the training of the baby's immune system.

8. The latest science indicates that infants born by Cesarean section could benefit from immediate skin-to-skin contact with the mother (which is now becoming possible in the operating theater) and breastfeeding where possible.

9. Dr. Dominguez-Bello from New York University is researching a swab-seeding technique in which an inoculum (microbial payload) of the mother's vaginal microbes is immediately introduced to a baby born by C-section, with the aim of introducing some of the microbes the baby would have acquired if he or she were born vaginally. Preliminary results using strict protocols indicate that this technique could partially restore a baby's microbiome.

Disclaimer: The description in this video clip of Dr. Dominguez-Bello talking about her research into swab-seeding is for general information only. An expectant woman should always consult with a physician before deciding whether or not to take any action relating to her pregnancy or if she has any concerns about her own or her baby's health.

CESAREAN SECTION
AND THE MICROBIOME

FIVE

What Is the Role of Bacteria in Training the Infant Immune System?

W e are under attack from foreign invaders pretty much every second of our lives, but the amazing thing is that we don't even notice the attack happening—at least not until we feel ill. This constant barrage is fought off by a tireless army known as the immune system.

The human immune system is a complex organization of organs, tissues, and cells, each of which has several functions to carry out in order to protect the whole human organism, or the "self." The role of the immune system is not only to fight off infection and disease but also to ensure that our organs and tissues exist in a healthy state of equilibrium, known as homeostasis.

While a properly functioning immune system will maintain a healthy body, one that isn't functioning properly could increase the risk of disease. It's essential, therefore, that this defense system is set up, or primed, correctly from the very beginning of life.

At birth a baby enters a world that is teeming with bacteria, viruses, fungi, and parasites. At this point, his or her immune system is not yet fully developed and is unable to tell which of these microbes may be harmful to the "self." It is of the utmost importance, then, that it be allowed to learn the difference between beneficial microbes and harmful ones as soon as possible.

Recent scientific discoveries have revealed that a major part of this process begins during labor and birth and that bacteria have a crucial role to play in its success.

How does the immune system work?

One way to picture the immune system is as a fortified city surrounded by walls and further protected by a defense force within. The walls represent the skin, as well as the mucous membranes that line the gastrointestinal (GI) and respiratory tracts, and are collectively known as epithelial barriers. They are the first lines of defense and ordinarily do a great job of repelling invaders on a daily basis. The skin and mucous membranes are so tightly packed with "friendly" microbes that harmful microbes, known as pathogens, find it difficult to find a place to settle and take hold. Once the immune system detects that a pathogen has arrived, epithelial cells secrete antimicrobial chemicals that inhibit the pathogen's growth. If the pathogen manages to make it into the GI tract, stomach acids and digestive enzymes will most likely kill it.

Should a pathogen succeed in penetrating the epithelial barriers, it will come up against the second line of defense: patrolling rapid-response soldiers known as phagocytes (such as macrophages). These will surround and absorb the invading cells before releasing digestive enzymes that will hopefully eat up the pathogen itself.

A sign that an immune response is taking place is inflammation, which is the expansion of narrow blood vessels in the infected tissue. We usually experience this as a tender swelling at the site of infection. Swelling means that our blood vessels are becoming "leaky" to allow other immune cells to rush to the area of infection.

If phagocytes fail to eradicate the invader, they will detain it until reinforcements arrive in the form of dendritic cells. The job of these cells is to escort the invader to the lymph nodes, which are home to specialist white blood cells known as lymphocytes, or B and T cells. The lymphocytes (B cells) will then study the pathogen and start to manufacture the antibody needed to immobilize that specific invader.

If a particularly hardy pathogen continues to survive, specialist T cells, known as cytotoxic cells, join the fight, often with the help of macrophage and dendritic cells. This combined attack is usually enough to ensure annihilation of the pathogen. The process is ongoing, complex, and somewhat remarkable!

What are innate and acquired immune responses?

What we have just described is a combination of innate and acquired (also known as adaptive) immune responses. The innate phagocytes (macrophages and dendritic cells) are the frontline "grunts." They move swiftly into battle and will always do the same thing when they engage the enemy. But they don't have the ability to learn from their experience.

The lymphocytes, on the other hand, learn from a pathogenic invasion. They use their "acquired" knowledge to remember what to do should that particular pathogen ever return in the future.

Innate and acquired responses are both powerful elements of the immune system, but they each also have their weaknesses. The effectiveness of an innate response is owing to its immediate action, but phagocytes attack without prejudice. This can result in collateral damage—that is, damage to surrounding healthy cells. An acquired immune response is effective because lymphocytes can produce the precise antibody to neutralize a specific threat. However, it can take up to a week or two to clone enough antibodies to do so. We feel ill or feverish when an acquired immune response is methodically cooking up the cellular soup to deal with an infection.

If we had only innate immunity, the heavy-handed response might cause too much collateral damage. If we had only acquired immunity, the delay in response might give a pathogen the necessary window of time to launch an infection that becomes too powerful to destroy.

The key to the success of this combined defense force is its ability to communicate between the different units that work together to determine the appropriate response for each invader. In some cases only one type of defender will mobilize to repel an attack, but in others it may take a coalition of all the units.

How do immune cells identify an intruder?

All cells have protein molecules on their surfaces known as antigens. These act as identity tags that express the genetic material, or DNA, contained within the cell, whether it be human, bacterial, or whatever else. Antigens allow the immune system to recognize cells that belong to the "self." They say, "This is me and I'm supposed to be here. Leave me alone." They also allow the immune system to recognize what is not part of the self: what is foreign and potentially harmful. And when these foreign antigens present themselves, the soldiers of the immune system are sent into battle.

Some T cells (T-helper cells) help B cells manufacture antibodies, which are Y-shaped proteins that bind to foreign antigens, thereby immobilizing them until they are destroyed by other immune cells such as phagocytes. T-helper cells also orchestrate the overall attack strategy. Other T cells (cytotoxic T cells) may kill an invader directly.

These are the main battle units of the immune system, but there are many, many more that have their own roles to play—too many to mention here.

When does the immune system develop?

This immunity army is not complete at birth; mostly it develops and matures all the way up until adolescence. The B and T cells continue learning throughout a lifetime, "acquiring" and banking intelligence on each pathogen, ready to repel any reappearances should they occur.

This acquired education is essential for the army to operate at full potential throughout our lives, but any disruption to the learning process can result in elements of the defense force becoming confused. When this happens, they may even turn their power against the self by attacking healthy cells, tissues, and organs, resulting in autoimmune and inflammatory disease.

What's particularly relevant to our discussion on birth is that recent scientific research has discovered that perhaps one of the most important elements of the immune system's education begins, as so eloquently described by Dr. Rodney Dietert, "in the narrow window that surrounds birth." It involves the interplay between the mother's microbes (coming from her vagina, her fecal matter, and her breast milk) and the baby's own naïve immune cells.

Why is a baby's immune system not complete at birth?

Since one of the main responsibilities of the immune system is to seek out and destroy foreign intruders (as identified by their DNA), we could ask how it is that a "foreign" fetus is allowed to grow within a human body. The cells of a fetus contain DNA from both the mother and the father. This means the antigens on the baby's cells will be identified as being at least partly foreign by the mother's immune system. Ordinarily this would be enough for the mother's immune system to launch an attack and reject the fetus.

Fortunately, there are some clever mechanisms by which the mother's immune system is tricked or informed that the fetus should be tolerated. The mother's T cells, which normally attack foreign cells, are prevented from targeting the fetus through a little "tinkering" with DNA.

The genes that would usually mobilize the mother's T cells to attack are switched off within the decidua (the structure surrounding the fetus and placenta). This results in the mother's T cells being denied access to the fetus.[1]

Additionally, during pregnancy, some of the mother's T-helper cells, called Th1 cells, are suppressed to further safeguard against rejection of the fetus. The baby's own Th1 cells remain undeveloped until after birth, which means the newborn baby is missing a vital part of his or her immune system.

While this may sound alarming, it is actually very important that the baby's immune system is suppressed in the first few weeks of life after birth.

Why does the baby's immune system need to be suppressed in the first weeks?

While the fetus is in utero, his or her immune system is fundamentally innate. The baby's key defense force is made up of the "grunts" (phagocytes) that will attack all the foreign cells they encounter. During the birth process, the baby will come into direct contact with millions of microbes in the birth canal, microbes that are of vital importance to the baby's development and future health. Since the baby's phagocytes will view these microbial cells as foreign, how is it that the baby's immune system doesn't destroy them?

Why doesn't the newborn's immune system attack the vaginal microbes?

The mother's vaginal bacteria (such as lactobacilli) and the mother's intestinal bacteria (such as bifidobacteria) need to be guaranteed safe passage through the baby's GI tract in order to seed the baby's gut microbiome. So how do they manage to get past the "grunts"?

Something needs to happen to suppress these killing machines. And it does.

During the latter stages of gestation, the developing baby produces immunosuppressive cells called CD71+ erythroid cells. These cells hold the baby's killer immune cells at bay during and after childbirth, allowing beneficial bacteria to take up residence in the gut.

We know that babies who are born prematurely often suffer from inflammation of the gut, an indication that an immune attack is taking place. This could suggest that prematurity has meant that the baby's immune system hasn't yet prepared any or enough CD71+ cells, so that when the mother's microbes arrive in the newborn's gut, the baby's innate killer cells launch an all-out attack instead of tolerating them. This attack can lead to an exceptionally nasty condition called necrotizing enterocolitis, which destroys intestinal tissues and can be fatal.[2]

By three weeks after birth, CD71+ cells have depleted in numbers, allowing the baby's immune system to kick in. And here, too, bacteria have a crucial role to play.

How does the immune system get its first lessons?

As we mentioned earlier, the grunts of the innate immune system do not have the ability to learn. It is the function of the acquired immune cells to learn what to attack and how. The first lessons the acquired cells receive will be from the mother's vaginal bacteria and then the bacteria that are contained within breast milk. These species of bacteria (including lactobacillus, bifidobacterium, and bacteroides) send out signals that start to interact with the baby's immune cells to begin educating the baby's immune system.

As Dr. Dietert explains, "Microbes, particularly those in the gut, help the immune system to mature, and they do so using a variety of different chemicals. Some of these chemicals are in solutions, some of them are on surfaces of cells and on microbes themselves, but they are all important in terms of the immune system maturing and understanding what is safe and what is not safe."

This process is called immune tolerance and determines which microbes the human body should tolerate and which it should attack. To return to our warfare analogy, the sentries on the walls are being taught the difference between friend and foe.

As we saw in chapter 3, breastfeeding provides more bacterial species and the complex sugars to feed them. But it also provides antibodies that help protect the baby while their own immune system is still being suppressed, to allow all those maternal microbes to get to work in the gut.

The breast-milk antibodies help protect the mucosal lining of the baby's gut, throat, and lungs, as well as defend against pathogenic bacteria and viruses. But these antibodies are only a temporary gift—they won't last forever, and eventually the baby's digestive system breaks them down and they are expelled from the baby's body.

How does a Cesarean section affect the education of the immune system?

If a baby is born by elective Cesarean section, his or her first encounter with the microbial world will not be with the mother's vaginal or intestinal bacteria. Instead it will be with airborne bacteria from the operating theater and bacteria, such as streptococcae and staphylococcus, from the skin of the medical staff and parents.

It will, therefore, be these species of bacteria that provide the initial education of the immune system. And this can come with adverse consequences.

Researchers hypothesize that without the "natural inoculum" from the mother at birth, the baby's immune system receives a kind of mal-education, which in turn could lead to immune malfunctions later in life.

In the CHILD pilot and follow-up, Professor Kozyrskyj found that babies born by Cesarean section had a different bacterial "footprint" than those babies born vaginally. "At 3 to 4 months of age, these infants have a lower abundance of this genus *Bacteroides* than vaginally delivered infants. We think that the genus *Bacteroides*, the one that we found to be lower in infants following Cesarean section delivery, may be one of the microbiota that's important in terms of this development of immune tolerance."

If the baby's immune system isn't trained correctly at the very beginning of life, there is a chance that it won't respond appropriately later in life.

According to Dr. Dietert, "The immune system is going to respond recklessly against what are not hazards from the environment. It is unable to decide what is safe and what is not safe, and so it errs on the side of everything is dangerous. But in doing so, it is going to produce things like allergies and allergic disease, inflammation in tissues, and autoimmune reactions. And that's what leads to later life disease."

While it may be possible to go some way to restore the balance of the microbiome, the problem, according to Dr. Dietert, is that in babies born by Cesarean section the initial education of the immune system does not happen correctly in the first place: "The microbes were missing during the critical period where it needed to be trained. So the immature, unbalanced immune system is going to go through that child, with the adult haphazardly responding to things and potentially creating disease in tissues."

Yet Professor Kozyrskyj's new findings again point to the longer-term benefits of breastfeeding. In the two hundred CHILD infants, depletion of bacteroides was no longer evident 1 year later among infants breastfed for at least three months.

What have we learned so far?

So far, we have discussed the importance of optimal seed and feed for the baby's microbiome in terms of preventing a baby from developing diseases later in life. As we've seen, if optimal seed-and-feed does not take place during and immediately after birth, there may be lifelong health consequences.

What we haven't considered is the ongoing impact—what about our children's children? If a child is born by Cesarean section, are there consequences for the generation that comes after them? Even beyond that generation, could increasing rates of Cesarean section have ramifications for the future of our species entirely? We'll be exploring these questions in the next chapter.

Here's a summary of the main points we've covered in this chapter:

1. The baby's immune system is not fully developed at birth.
2. During pregnancy, immune helper cells known as Th1 cells are suppressed in the mother to stop them attacking and rejecting the fetus. This means that the fetus's own Th1 cells are held back from developing, too.
3. Up until three weeks after birth, CD71+ cells suppress the baby's immune system, allowing bacteria to colonize the baby's gut microbiome.
4. With vaginal birth, bacteria from the mother's vaginal and intestinal microbiome arrive and colonize the baby's gut. These bacteria kick-start the continued development of the infant immune system.
5. The bacteria train the immune system to identify what is friend and what is foe. This means the immune system is educated as to which microbes are beneficial and should be tolerated, and which microbes are harmful and should be attacked.
6. All the way through gestation, birth, and infancy, certain things are supposed to happen at certain

times. These events happen only once. If they don't happen within their designated window (and Cesarean section and formula feeding have the potential to make the events miss their window), the baby may develop a malfunctioning immune system.

THE IMMUNE SYSTEM
AND THE MICROBIOME

How Is the Mother's Microbiome Passed on to Future Generations?

If an infant is born vaginally, the baby's gut microbiome resembles the mother's gut microbiome. The two are not exactly the same because of all the other microbiota the baby is exposed to during birth—but they are a close match. This means that a baby born through the vagina has many of the same microbial strains in its gut as its mother's because the mother has passed them on to her baby.

Put another way: A child born vaginally has almost the same microbial footprint as its mother.

This is a critical transgenerational aspect of the human microbiome —it is transgenerational because the microbes are passed from one generation to the next, and so on. As Dr. Maria Gloria Dominguez-Bello says, "It's like a maternal heritage."

If a mother passes her gut microbes to her daughter during a vaginal birth, the daughter (if she goes on to have children and to give birth to them vaginally) will pass on her microbial footprint to her own babies. In other words if all the generations are born vaginally, grandchildren will have the same types of microbes as their maternal grandmother, and so on stretching back through the female line to the very roots of that family tree. The key to this incredible maternal lineage is that each baby passes through her mother's birth canal, thereby receiving an "inoculum" of vaginal microbes, and each baby is also likely to come into contact with the mother's fecal matter, thereby exposing the infant to the mother's gastrointestinal microbes.

According to Dr. Rodney Dietert, "We know that the baby's gut reflects precisely the microbes that are in the mother's gut. They are passed along, much as a mother would pass chromosomes on to the baby."

What does this mean for Cesarean section babies?

Because during a Cesarean section the baby is pulled directly from the womb into the air of the operating theater, rather than being born through the vagina, the opportunity for him or her to acquire the mother's vaginal or intestinal microbes is dramatically reduced. Because the Cesarean section means that the environmental microbes seed the baby's gut, the baby could start life with a different set of bacteria from the mother. In other words, the mother and the baby have a different microbial footprint from each other, and the chain of microbial lineage is broken.

If a Cesarean section–born baby girl goes on to have children of her own, even if she gives birth to them vaginally, those children will have different bacterial species in their gut microbiome from those in the gut of their maternal grandmother—and so on, back through time.

If any of the bacterial species that are missing from the maternal heritage are "keystone species," as described by Dr. Martin Blaser (see page 34), the lack of those species can persist through future generations. This means that, unless the child can acquire those powerful keystone species elsewhere, future generations might be more susceptible to disease. A Cesarean section in one family member, therefore, may have consequences for the health of future generations of that family.

What would we have done differently?

As parents of a child born by C-section (albeit emergency C-section), we live with the uncomfortable reality that our daughter's gut microbiome could be inherently different from her mother's and her maternal grandmother's and her maternal great-grandmother's.

Perhaps there's nothing medically anyone could have done differently. Certainly, though, if we'd known back then about the seeding of the baby's microbiome, there are certain procedures we'd have chosen to implement at the time of our daughter's emergency entry into the world. If only we had known that it is possible for the mother to have skin-to-skin contact with the baby in the operating theater within moments of the birth. If only we had known that even a little bit of formula milk could potentially interfere with microbial maturation. And if only we had known about Dr. Dominguez-Bello's swab-seeding technique to partially restore the microbiome of C-section babies.

Obviously, the life of mother and baby comes first. Who's to say whether in our emergency situation, we would have remembered any plan to help seed the baby's microbiome with the mother's vaginal and intestinal microbes? But if we'd had then the awareness we have now, and the opportunity, we would have done whatever we could to make sure Toni's vaginal and intestinal microbes seeded our daughter's gut microbiome.

If Toni's microbiome was screened during pregnancy, and if there were no concerns about the health of her vaginal microbiome, and if there had been no sterile gauze handy, perhaps we would even have gone as far as "DIY swab-seeding" our daughter's gut microbiome right there on the operating table as soon as the

C-section was completed. By this we mean we could have swiped some vaginal fluids and wiped them over our daughter's mouth, face, and body as soon as feasibly possible once she was out of the womb. Perhaps even wiping her with some fecal matter, too. Knowing then what we know now, would we have done it? Quite possibly, yes.

Let's be clear. We are not recommending this DIY procedure, or encouraging anyone else to try it. Apart from anything else, as we've already said, Dr. Dominguez-Bello's swab-seeding research is still ongoing. However, as parents who fully understand that every mother has the human right enshrined in law (at least in Europe) to decide the circumstances of her child's birth,[1] we would have taken full responsibility for the future health of our child, and we would have signed any hospital waiver or release form we had to—if we had had the opportunity (or the knowledge).

This feeling of "if only we had known" has become a positive driving force. It galvanizes us to make films and to write this book to raise awareness of this very subject. We are driven to communicate this information so that other parents, doctors and midwives might take all the research into this subject into account—particularly if a birthing mother needs a Cesarean section, or if parents are considering an elective Cesarean section with no medical indication. What might be right for us might not be right for other parents. There's not a good or bad choice—there's just a choice or set of choices that are right for individual people in a particular situation based on the information available to them at that time. For us what's important is that every parent has access to, and support for, fully informed choice.

What is epigenetics?

So far we've talked about the relatively straightforward notion of maternal microbial lineage. Epigenetics is where heredity gets complicated. We're about to discuss another microscopic effect that could also be happening during childbirth.

114

Put very simply, epigenetics involves the switching on and switching off of our genes, the caretakers of our traits and appearance, our tendencies, characteristics, likes and dislikes, and our predisposition to certain genetic diseases—and every other aspect of our uniqueness. We inherit our genes in our chromosomes from our parents. According to the UK's Science Museum website, we're born with about 24,000 genes: "This is a few more than a chimpanzee or a mouse, but nothing spectacular."[2]

Furthermore, over the course of our lives, our original 24,000 or so genes will never change. We are born with the same genes with which we die. Our gene structure never changes either. The expression of our genes, on the other hand, does change over the course of our lives.

Imagine that every person is born with thousands of internal light switches that can be either on or off. Something may trigger one light switch to turn on, while another is turned off—thousands of times over, again and again. This is exactly what happens to genes—a great imaginary epigenetic finger arrives and turns on or off our different genes, many at the same time. Scientists say that when a gene is turned on, that gene is "expressed"; when it is turned off, scientists say that a gene is "not expressed."

So, what causes the genetic switches to turn on or off? Environmental triggers ranging from exposure to chemical pollutants, changes to diet and lifestyle, even temperature changes and other external stresses can, according to a paper published in *Nature Reviews Genetics*, "indeed have long-lasting effects on development, metabolism, and health, sometimes even in subsequent generations."[3]

When we talk about someone having a "genetic predisposition" toward something, we mean that they have inherited a potential trait through their bloodline (perhaps a predisposition to a particular genetic disease or behavior), but that doesn't necessarily mean that gene will be expressed for that person—that might depend upon whether the person is exposed to the particular trigger for that gene to be expressed. This triggering process is the field of epigenetics.

Dr. Jacquelyn Taylor has another way to explain epigenetics: "We know that a lot of diseases are heritable in nature and are passed down from the parents. If your parent has a risk factor for a particular disease, you could have that risk factor based on what's passed down to you from that particular parent. Now, when you're looking at epigenetics, we're not talking about changes in that particular gene, we're talking about changes 'above the genome' that can cause the gene to express in a different way."

For example, you might have inherited a gene mutation associated with developing a certain disease. By mutation, we mean that there's an alteration in the DNA sequence that makes up a gene that is different from most other people. That doesn't mean you'll definitely develop that disease later in life, but you might be more "genetically predisposed" to it than someone who doesn't have that gene mutation. But whether or not you develop that disease may come down to the existence or absence of the trigger that forces the gene mutation to turn on or turn off. Remember that triggers are generally environmental—they are outside you. Something *around* you may determine whether or not you go on to develop that particular disease.[4]

Take a set of identical human twins. They have exactly the same genome, which means their sets of roughly 24,000 genes are absolutely identical. However, one twin might grow up to develop a certain noncommunicable disease (for example, depression, diabetes, or breast cancer), while the other twin does not. According to Professor Tim Spector, head of twin research at King's College London, "In each case we have found genes that are switched on in one twin and switched off in the other twin. This often determines whether or not they are likely to get a disease."[5] So an environmental trigger might switch on the gene for depression, diabetes, or breast cancer in one twin, but the second twin might never have been exposed to that crucial environmental trigger or might have been exposed to it in a different way, in which case, for the second twin, the gene for the disease never activates.

In experiments with mice that carry the agouti gene (a gene that determines if the animal is banded or solid in color and also makes these mice very sensitive to epigenetic changes), identical twins with identical genes can look extremely different in color and size. In twin mice, one twin, for whom the agouti gene was switched on, grew up to be blondish yellow and obese; the other twin, whose agouti gene remained off, grew up to be brown and healthy in size. In 1994, Dr. David M. J. Duhl and his colleagues at Stanford University School of Medicine, United States, published research in *Nature Genetics* showing that mice can be "genetically identical, epigenetically different."[6] Something in the environment of the yellow, obese mouse flicked on the agouti gene switch. A follow-up article in *Nature Education* in 2008 discussed the wider relevance of this research, including possible environmental triggers for the agouti mouse experiment. One suspected trigger was bisphenol A, a chemical commonly found in plastic bottles.[7]

Where does all this fit into childbirth? Well, scientists are now beginning to consider whether childbirth itself might be one of the environmental triggers that causes certain genes to switch on or off.

What's the evidence for epigenetic changes in childbirth?

Some scientists are starting to hypothesize that while a baby is developing within the safety of the mother's womb, certain genes related to life outside the womb remain switched off. Then, when the baby is vaginally born, the natural stresses and pressures of the journey through the birth canal, particularly as the infant is squeezed through the mother's pelvis, could turn out to be critical environmental triggers that switch on the genes that the baby needs for a healthy life outside the womb. The cascade of hormones the mother releases during a vaginal birth, or even the baby's own release of stress hormones, may also be triggers for gene activation.

With vaginal childbirth, the process of physiological labor and birth could be the signal that, after nine long months of waiting, the genes associated with life outside the womb urgently need to be turned on. Though lots more research is needed for any certainty, some researchers are starting to speculate that the genes turned on through vaginal childbirth are associated with the baby's immune or metabolic systems. At the same time, the genes the baby has needed up until now for survival within the womb switch off. Again, some researchers speculate the genes that might urgently need to be turned off at birth are those associated with dependency on the mother's immune or metabolic system, or perhaps any genes associated with the placenta or the umbilical cord. For example, genes associated with receiving oxygen and nutrients through the cord might urgently need to turn off for the sake of the baby's lifelong health.

Members of the Epigenetic Impact in Childbirth (EPIIC) international research group, among others, are presently looking into the hypothesis that childbirth is an epigenetic event.[8] According to one of the founders of EPIIC, Professor Hannah Dahlen, "Birth has got to be one of the most cataclysmic events in a human's life. There is no other event that is so finely tuned, involves so many hormones, is so fundamental in shifting you from one being to another being, that you can't look at that event without saying, 'This is potentially really important for shaping us genetically.'"

How does a Cesarean section affect epigenetics?

As we've already discussed, there's a big difference between a Cesarean section with labor, and one without. If a mother is in active labor before undergoing Cesarean section, the baby may have experienced some of the physical sensations and hormone releases associated with vaginal childbirth. This, in turn, may have triggered some or all of the epigenetic changes that would have occurred if a baby is actually born vaginally.

If, however, a Cesarean section occurs before the mother is in active labor, or if it is an elective Cesarean section, the baby is likely not to have experienced the physical stresses of being squeezed through the mother's pelvis, nor the hormonal and stress influences of a vaginal birth. In essence, the baby might not be physically or psychologically quite ready to be born.

Think about when an alarm wakes you in the morning, signaling that it's time to get up and get going, according to your usual, well-organized, relaxed routine. The alarm is like the hormones and stresses of vaginal birth—preparing you for the day, just as the hormones and so on prepare the baby for life. If the alarm didn't go off, you might keep on sleeping. When eventually you did wake up, you'd most likely wake up with a start. You'd probably be groggy and confused, and a little panicked. The sequence of events of your waking is completely different from that of waking with the alarm. The outcome for the start of your day—and perhaps persistently into your day—is changed. Similarly, the Cesarean section baby without labor gets no alarm to say that it's time to wake up into the outside world. Instead, the unexpected sequence of birth events could very well cause confusion and stress—and completely change the sequence of switching on and off of the baby's genes, perhaps even leading to a different lifetime outcome.

We're still at the very early stages of knowing the exact relationship between vaginal birth and epigenetics, and indeed the relationship between Cesarean section and epigenetics. However, it is a very real possibility that a surgical birth could interfere with certain epigenetic processes. The result of this could be that the right genes do not switch on at the right time, while others do not switch off at the right time. This interruption in natural epigenetic processes could interfere with the optimal functioning of the human body, with potentially serious consequences for human health further down the line.

What does the research into the epigenetics of childbirth suggest?

Although there is some research that has looked at epigenetic changes during pregnancy, and other research that has looked at epigenetic changes post-birth, there have been very few large-scale long-term studies exploring potential epigenetic changes happening during childbirth itself.

Dr. Taylor is also a member of the EPIIC research team, and she believes there's a very simple explanation for this lack of research: "I think epigenetics is just a very new area of research for the perinatal period. In fact, much work has been conducted in adult populations, and my laboratory is currently investigating early life epigenetic changes in children as young as 3 years of age. However, I think the short period of time that we're focusing on here, the time from labor to delivery, has been missed."

Some research has been done. According to Professor Dahlen, "There are a couple of studies that have come out showing that there are methylations, or changes seen genetically, with Cesarean section. The studies have shown that potentially Cesarean sections are silencing our genes or parts of our genes that are really important for the immune system."

One recent study looks at potential epigenetic changes in the baby's stem cells associated with Cesarean sections.[9] "A possible interpretation is that mode of delivery affects the epigenetic state of neonatal hematopoietic stem cells. Given the functional relevance indicated, our findings may have important implications for health and disease in later life."

If common interventions in childbirth do indeed cause epigenetic changes, and if these changes are linked with certain health conditions later in life, the hope is that researchers might be able to identify correctly which medical interventions are associated with which particular epigenetic changes. If they can do this, they could pave the way for other researchers to find ways to prevent or treat these health conditions in the future.

According to Dr. Taylor, "We expect to find—and this is the hypothesis—that there are changes that occur to the epigenome during the birth process. If we're right and if we can identify those changes, then that can provide the foundation for us to improve health outcomes via personalized interventions." She adds, "I think the consequences for mankind are that we can hopefully reduce the risks for development of common and chronic diseases in the long term, such as asthma, diabetes, and hypertension."

What are the transgenerational effects of epigenetics in childbirth?

Some epigenetic changes are temporary; other epigenetic changes are permanent. The permanent changes are particularly concerning because some of these changes could even be passed on to the next generation. This means that, potentially, the epigenetic effects related to a single birth can be multigenerational.

To explain this in another way, the tendency for a certain gene to be switched on or off can be passed on to the next generation. According to Professor Dahlen, "Epigenetic changes that silence parts of your genes can then be passed on to the next generation. This means the whole process of birth has a potential impact on reshaping us genetically in future generations."

To express the same idea in yet another way, an epigenetic change during childbirth could be related to a positive health outcome later in life—for example, a child might grow up without developing asthma or other atopic conditions. Or an epigenetic change in childbirth could be related to a negative health outcome later in life—for example, the switching on of a gene associated with the child developing a particular disease, such as asthma. Perhaps it might not. Whether they are connected with positive or negative health outcomes, whatever epigenetic changes are occurring, they could be passed on to the next generation.

Up until now most research into the transgenerational effects of epigenetics has been in animals—and not all animal species behave in the same way. Because each species is unique, including humans, we must treat research using animals with caution. Nonetheless there is increasing evidence to support the hypothesis that epigenetic changes are passed down through generations. According to Professor Dahlen, "We know that if your grandmother smoked, there are changes seen in girl babies two generations later. Absolutely we know that these epigenetic changes can go on and be intergenerational."

Recent research indicates that epigenetic changes could have a multigenerational ripple effect, finding their way into the genetic makeup of descendants many generations, even centuries later. According to Dr. Dietert, "Effects during pregnancy can extend not just to the offspring from the fetus, but to that child's child, to the grandchildren. There is some evidence to suggest it might affect generations beyond that. So we see a minimum of a 100-year effect in terms of health risk."

This means, potentially, epigenetic changes during childbirth could impact the health of people 100 years into the future, and even beyond. In other words what happens in a single birth today could impact the health of that child's great-great-grandchildren, even great-great-great-grandchildren.

The last word on the ramifications of epigenetic research in childbirth goes to Dr. Taylor: "The sooner we can identify what epigenetic changes are occurring, and if there are truly risks for negative health outcomes down the line, we can make a significant impact on health for all mankind." Simply, we need more research.

Here's a summary of the main points we've covered in this chapter:

1. The microbial profile of a baby born vaginally resembles the microbial profile of its mother. A mother passes down her unique set of microbes to the next generation, if she has a vaginal delivery.
2. A Cesarean section potentially breaks the chain of "maternal lineage"—the mother most likely does

122

not pass on a complete set of her unique microbes to her child. This could mean the child fails to inherit certain critical keystone species of bacteria that help protect against disease.

3. Epigenetics is the study of the triggers that switch a gene on or off. Whether a gene in your body is expressed (switched on) or not may depend upon an environmental trigger—something outside of you that makes certain genes in your body switch on or off.

4. Scientists hypothesize that the physiological processes associated with vaginal childbirth could themselves be particular environmental triggers designed to switch certain genes on or off. A Cesarean section provides an altogether different physiological process, perhaps affecting the activation or deactivation of certain genes. This might itself have long-term health consequences for the baby.

5. Epigenetics also has the potential for transgenerational effects—meaning that any epigenetic changes during childbirth may be passed down to future generations.

EPIGENETICS EXPLAINED

SEVEN

Is There a Link Between Cesarean Sections and Disease?

In terms of the short-term health risks for babies born by Cesarean section, a baby born surgically may experience respiratory problems or breathing difficulties, particularly if the Cesarean section occurs before 38 weeks' gestation. At 38 weeks a pregnancy is considered full term (defined as being between 38 and 42 weeks' pregnant).

An expert in this field is Dr. Neena Modi, who has clinical duties as a neonatologist at a London hospital. "I look after babies in a neonatal unit, and it's certainly true that we do see a substantial number of babies who are admitted because they've experienced breathing problems after being born by Cesarean section." According to Professor Modi, "We know if you're born by a pre-labor Cesarean section, at around 37 weeks of gestation, about 1 in 10 of those babies will develop respiratory problems and will need to be admitted to a neonatal unit. However, if you wait until 38 weeks, that risk has fallen to about 5 or 6 percent. And if you wait to about 39 weeks, it's fallen even further, to about 2 to 3 percent."

While the fetus develops inside the womb, its lungs are filled with fluid, and oxygen comes from the mother's placenta. However, before the newborn takes his or her first breath in the outside world, they need to clear the fluid from the lungs and produce surfactant, a special lubricant that allows the lungs to draw in air as soon as the baby is born. This process, as well as changes in blood flow, energy metabolism, and temperature control, are some of the critical physiological processes occurring as the fetus transitions to newborn. These adaptations usually happen naturally through the stresses, pressures, hormone releases, and other experiences associated with vaginal birth at term. However, they don't necessarily happen—or at least happen in the same way—for a baby born before 38 weeks or by Cesarean section. There may be "abnormalities" in the physiological transition from fetus to newborn. As described in one scientific paper, "Abnormalities in adaptation are frequently found following preterm birth or delivery by Cesarean section at term, and many of these infants will need delivery room resuscitation to assist in this transition."[1]

As well as being more likely to develop mild breathing difficulties and respiratory problems, babies born surgically are—in the short term—more likely to have low blood-sugar levels, be less alert, and be sleepier, and they may find it more difficult to feed successfully.

Our own daughter needed to be resuscitated immediately following her C-section birth at term (Toni was 40 weeks plus 4 days). Our baby was very drowsy for the first 24 hours; she had low blood-sugar levels; and we had difficulties establishing breastfeeding. In fact we ticked all the boxes for the recognized, well-documented short-term problems associated with being born by C-section.

But what about potential long-term problems?

What are the long-term health risks for a Cesarean section baby?

Over the past few years, a number of large-scale epidemiological studies have suggested a strong link between Cesarean section birth and an increased risk of developing certain diseases later in life.

According to Dr. Martin Blaser, "There are more and more studies showing that children born by C-section have a later risk for a variety of problems, ranging from diabetes to obesity to allergic diseases like celiac disease."

Strong epidemiological evidence shows that a child born by Cesarean section is at increased risk of developing a number of chronic health conditions:[2]

- Approximately 20 percent increased risk of developing asthma[3]
- Approximately 20 percent increased risk of developing type 1 diabetes (also known as juvenile diabetes)
- Approximately 20 percent increased risk of becoming obese or overweight later in life
- Approximately 15 percent increased risk of developing celiac disease (an autoimmune inflammatory condition associated with intolerance to gluten), though one study found an 80 percent increased risk[4]

Statistics are perhaps easier to understand when you have actual numbers. In one recent study by Dr. Jan Blustein of New York University School of Medicine and Dr. Jianmen Liu of Peking University published in the *BMJ* in June 2015, the researchers found that in the United States, the overall childhood asthma rate is 8.4 percent, but this jumps to 9.5 percent if a baby is born by Cesarean section. Type 1 diabetes occurs in 1.79 per 1,000 babies born vaginally, compared with 2.13 babies born by Cesarean section. In terms of obesity, they found that the obesity rate among children born vaginally is 15.8 percent compared with 19.4 percent of children born by Cesarean section.[5]

All these conditions are related to the immune system, a point made by Dr. Maria Gloria Dominguez-Bello. "These studies really show dramatic increases in conditions and diseases that have to do with high inflammation and bad reaction of the immune system."

Of course, not all Cesarean section births will result in babies with these conditions, or even a combination of them. Some

Cesarean section children might not experience any of these conditions at all. Nonetheless, statistically, children born by Cesarean section are at significantly increased risk of developing asthma, type 1 diabetes, or celiac disease, and of becoming obese or overweight later in life.

Is an association the same as a cause?

There is an important point to make here: ASSOCIATION DOES NOT EQUAL CAUSATION. Just because research shows that Cesarean sections are connected to or associated with an increased risk of a child developing certain health conditions, it does not mean Cesarean sections actually cause those conditions.

For example, in relation to obesity, Dr. Neena Modi explains there could be other factors at work: "You might find that the increased risk of being overweight or obese if you're born by Cesarean section is actually because your mum was overweight. The risk isn't anything to do with how you were born, but it's to do with the genes you inherited, or it might be due to the environment in which you were brought up. You can see that we have to be really careful about whether or not actual mode of delivery causes these problems."

There may be strong associations between obesity and Cesarean section birth (or between any of the other health conditions and Cesarean section birth), but we haven't yet determined the exact nature of that relationship.

What are the ethical limitations of research?

Scientists face a big problem in how to obtain proof that birth interventions, including Cesarean sections, could potentially be "causing" long-term health problems—it's extremely difficult to do causal experiments on humans, and especially difficult if you're looking to do causal experiments with women during labor and birth.

The next best way to show causation is through use of animal models, as Dr. Dominguez-Bello explains: "When I say 'epidemiological evidence shows,' that means no causation is demonstrated through these studies. However, the way to show it is going to the lab using, for example, animal models. Then trying to see if you can cause a disease. Because you cannot do that with humans, animal studies are really the only way to demonstrate causation."

Dr. Blaser also notes the difficulties in researching causal studies in humans: "We can't subject children to the kinds of exclusions that we need for causal relationships. That's why we study animals. We can build up a body of knowledge with correlation studies where the exposure precedes the effect. We can show temporal relationships."

Longitudinal studies can help us with establishing cause and effect, too—although they aren't able to precisely elucidate causality. A longitudinal study involves following, investigating, and assessing a group of people over a certain period of time, using certain lifestyle, health, and hereditary markers to measure the participants' risk for and appearance of certain diseases. For example, you might select a group of Cesarean section–born babies and a comparable group of babies born vaginally (you'd choose babies born into roughly the same culture, environment, and family unit in order to minimize the risk of skewing your results), then follow the health of both groups over a number of years—often decades.

Are noncommunicable diseases the tip of the iceberg?

Asthma, type 1 diabetes, celiac disease, and obesity are all examples of noncommunicable diseases (NCDs). They are sometimes referred to as "chronic diseases." NCDs are conditions that cannot be passed from one person to another—they are not transmissible or infectious. You can't catch them as you would a common cold, for example.

However, Dr. Dietert believes these NCDs are "just the tip of the iceberg" when it comes to the health consequences of Cesarean sections on those babies. Developing one of these conditions in early life could pave the way for developing other related conditions much later on. "We know that noncommunicable diseases are interlinked. That is, if you have one, particularly as a child, your chances of having another and maybe a third later in life are very much increased."

According to Dr. Dietert, if a child develops asthma when they are growing up, that child "has a greater risk of being obese, of having behavioral issues, or eventually developing lung cancer than children who don't have asthma."

Taking asthma as an example, if a baby is born by Cesarean section, they are at around a 20 percent increased risk of developing asthma later in life. According to Dr. Dietert, having asthma early in life increases the risk of developing other conditions later in life, including becoming overweight or obese, and having anxiety or depression. Research suggests that asthma is also a risk factor for developing lung cancer. This projected trajectory of disease is true of the whole population, not just those born by Cesarean section. However, if being born by Cesarean section makes an infant more susceptible to developing asthma later in life, it stands to reason that the more Cesarean sections there are, the higher the incidence of asthma. And the more Cesarean sections there are, the higher the incidence of the other serious physical and mental health conditions associated with asthma.[6]

In his research, Dr. Dietert found a similar trajectory with type 1 diabetes. A child that develops type 1 diabetes is at increased risk of developing celiac disease, depression, renal failure, cardiovascular disease, hypertension, and many autoimmune conditions, as well as becoming overweight or obese. It's the same when we look at celiac disease. The development of celiac disease early in life is associated with increased risk of many other conditions later on, including cardiovascular disease, psoriasis, and osteoporosis.

To be clear, the trajectory of future conditions associated with type 1 diabetes and celiac disease is true of the whole population,

regardless of mode of birth—it's just that as we increase the numbers of Cesarean section births, so we increase the incidence of long-term disease.

The trajectory of diseases across a lifetime connected with asthma, type 1 diabetes, and celiac disease is explained in more detail in a video featuring Dr. Dietert (the link to which is at the end of this chapter). In this video, a still from which appears above, Dr. Dietert draws out and explains a "Tree of Disease" associated with being born by Cesarean section.

Dr. Dietert connects this increased susceptibility to immunological disorders associated with being born by Cesarean section to the immune system not maturing correctly at birth: "If we don't complete the self at the window when it is ideal or optimized, a system like the immune system doesn't mature correctly. Among other things, it grows up and matures not as effectively as it would. So the responses are different, your likelihood of autoreactivity or autoimmunity, for example, of allergic diseases is dramatically increased."

According to Dr. Dietert, a child who does not "complete the self" by being exposed to the right bacteria in the narrow window that surrounds birth could be more susceptible to a whole range of serious health conditions, including heart disease, obesity, cardiovascular disease, asthma, allergies, and autoimmune conditions (such as type 1 diabetes, lupus, and rheumatoid arthritis). These are all conditions that are rapidly increasing in developed and developing nations alike.

If Cesarean sections continue to become more prevalent, according to Dr. Dietert, "You're going to see this play out with a large percentage of those babies having asthma or allergic diseases, food allergies, type 1 diabetes, type 2 diabetes, cardiovascular disease, and autoimmune spectrum disorders as well. We will have more children with dysfunctional immune systems; we will have more inflammation in tissues, we will have more diseases occurring each decade as the child ages."

How do gut microbes relate to brain development?

Some research suggests that having a Cesarean section could affect infant brain development. This is because there may be a direct link between what happens in the baby's gut during and immediately after birth and the development of that baby's brain.

A number of neural behavioral disorders have been linked to an altered infant gut microbiome. According to Dr. Dietert, "We know when the baby is not seeded correctly, the baby is incomplete, those microbes that are present produce a particular spectrum of by-products and those by-products can influence neural behavior. They influence the baby and the infant's behavior. And there's a great deal of interest about how this may come into play relative to conditions such as autism."

The possible connection between mode of birth and a complex condition such as autism is very sensitive. Just to reiterate, the research is a still at a very early stage, and it will be many years before we uncover the whole picture. However, the first few pieces of the puzzle suggest it would be worth researchers trying to find some of the missing connections to complete the picture one way or another in the future.

Dr. Dietert explains, "What is known is that a portion of children with autism spectrum disorders have altered gut microbiota. They are not as diverse as is found in other children. They also have leaky gut, which means that the gut integrity is not what it should be. And it turns out that those microbes are making by-products that we know influence the brain, influence the nervous system tissues, in a way that controls behaviour. So it appears that there is an important link between the gut microbes and neural behaviour."

What plausible explanations are there for the apparent long-term health implications of Cesarean sections?

As we've seen, at the moment there seem to be two leading explanations for the association between Cesarean section and the baby's increased risk of developing various chronic conditions. There may be others, but right now we have two main contenders under investigation all over the world: altered gut microbiome and epigenetics.

What does it mean to have an altered gut microbiome?

Babies born by Cesarean sections have altered gut microbiomes—it's not the same as the microbiome via vaginal birth. This could impact the training of the baby's immune system and their metabolic function.

According to Dr. Dominguez-Bello, "We have an epidemiological association of C-section-born babies having high risk of certain diseases that are related to bad immune functioning. On the other hand, we have proved that babies born by C-sections do not get the right set of bacteria, and they start the immune system education with the wrong set of bacteria. So if we put those two things together, we have a very strong basis for a hypothesis that there is a relationship between not having the right set of bacteria educating the immune system and later health consequences."

She goes on to say, "When the baby's an adult, the immune system starts attacking the wrong antigens. Like reacting against gluten. Like dermatitis as a response to antigens that are not really harmful at all. These are the kind of diseases that we see now in the young generations like never before. These are rocketing diseases. Malfunction of the immune system, which is reacting as if it were fighting against a pathogen when there is no pathogen."

What about epigenetic changes?

Alternatively, the long-term health risks to a baby born by Cesarean section could be related to epigenetic changes in childbirth. According to Dr. Modi, "As a potential mechanism, it might be that a baby needs to be exposed to the stresses and hormones that come with normal labor to turn on certain genes. And that if these genes aren't turned on, that sets us on a health trajectory towards these long-term health problems."

Is there a third way?

Or, it could be a bit of both—that is, alterations in the baby's gut microbiome combined with epigenetic changes. Perhaps there's a complex interplay between genes and microbes. Perhaps there's another explanation altogether—only time, money, and further research will tell. This might take a decade, or perhaps even longer.

As Dr. Modi says, "As a scientist, what I try to do is look at what's plausible and do the research to see whether or not that can be substantiated. I think that's the next 10 years of research endeavor because we can't go on forever hypothesizing about the possibility that there is a causal connection between Cesarean sections and long-term health problems. We've really got to start to address it as a distinct possibility."

So, until we have definitive answers from further research in the future, what can we say for sure right now?

Dr. Dietert offers an answer to this question: "Over the past 20 to 30 years, we've seen dramatic increases in childhood asthma, type 1 diabetes, celiac disease, and childhood obesity. We've also seen increases in Cesarean section delivery. Does Cesarean section cause these conditions? No. What Cesarean section does is not allow the baby to be seeded with the microbes. The immune system doesn't mature. And the metabolism changes. It's the immune dysfunction and the changes in metabolism that we now know contribute to those diseases and conditions."

Here's a summary of the main points we've covered in this chapter:

1. Strong epidemiological evidence indicates that Cesarean section significantly increases the risk of a baby developing serious health conditions later in life, including asthma, type 1 diabetes, or celiac disease, or becoming overweight or obese.
2. Not all babies born by Cesarean section will develop one of these autoimmune conditions—they are simply more susceptible to developing one or more of them in later life.
3. These conditions could just be the tip of the iceberg. If a baby develops one of these noncommunicable diseases early in life, they are at increased risk of developing more noncommunicable diseases later on, opening up a gateway to future disease lasting a lifetime.
4. Associated conditions include: heart disease, cardiovascular disease, bowel disorders, autoimmune conditions, mental health conditions, and even some cancers.

5. As there's a gut–brain connection, a number of neural behavioral disorders have been linked to an altered infant microbiome connected with being born by Cesarean section.

6. Currently, there are two possible theories to explain the connection between Cesarean section and later disease: First, babies born by Cesarean section have an altered microbiome (when compared with vaginally born babies); second, there may be epigenetic changes connected with a surgical birth. It might be one or a combination of these theories, or another reason altogether. Only time and more research will tell.

DR. DIETERT'S "TREE OF DISEASE"

EIGHT

What Is the Impact on Humanity as a Whole?

Noncommunicable disease (NCD) is a serious global health problem in both developed and developing nations. According to the latest statistics from the Centers for Disease Control and Prevention (CDC), NCDs such as cardiovascular disease, chronic respiratory disease, diabetes, and cancer account for 75 percent of all deaths worldwide.[1] Collectively, then, NCDs form the world's biggest killer. According to World Health Organization (WHO) statistics, nearly three-quarters of these deaths now occur in low- and middle-income countries.[2]

Treatment and prevention of these diseases are currently very high on governmental and intergovernmental agendas. In 2011, world leaders gathered for a high-level meeting at the United Nations (UN) to talk about NCDs as an urgent health crisis.[3] It was only the second time in the UN's history that international governments met to discuss a health topic (the first occasion was to discuss HIV/AIDS).

Ban Ki-Moon, the UN secretary general, said at the time that the summit "is our chance to broker an international commitment that puts noncommunicable diseases high on the development agenda, where they belong."[4]

One of the main strategies employed by the WHO and global health systems to reduce NCD levels is to limit the four risk factors traditionally associated with NCD: tobacco use, harmful use of alcohol, physical inactivity, and unhealthy diet.[5] These four risk factors can produce physiological changes, such as increased blood pressure, blood-sugar levels, and fat levels, which in turn contribute to disease.

There are lots of government initiatives aimed at encouraging people to give up smoking and excessive drinking and to improve diet and lifestyle. This is to be applauded and supported, but is it enough?

Perhaps international governments, and organizations such as the WHO and the UN, are looking in the wrong direction when it comes to ways in which to reduce the incidence of NCDs. Perhaps they all need to turn their attention and funding toward birth. Devising and implementing more programs to promote and support vaginal birth where possible, and to reduce the rate of Cesarean sections, could help stem the rising tide of chronic disease.

Imagine that we are all on board a giant rocket ship hurtling through space. We are all busy getting on with our lives, completely unaware that the rocket ship is heading for disaster—a collision with a far-off planet. Then, the countdown begins. Alarms sound and panic sets in. Leaders of the different nationalities on board come together to think of ways to change the rocket's trajectory, or to slow down the rocket to give us more time before impact. They roll out plans to get everybody on board eating healthily and exercising, with the aim of lightening the ship in a bid to slow it down. Then something dreadful happens. Someone discovers there's an ingredient in the rocket fuel that's causing the ship to accelerate. No one realized because no one checked. There's still time to change the rocket fuel to avert disaster, but the world leaders have to act now.

Isn't it time for scientists, researchers, funders, governments, world leaders, and international bodies to turn their attention to birth, to find ways to slow down our trajectory toward a future full of chronic disease?

What does NCD mean for the global economy?

In September 2011, the World Economic Forum and the Harvard T. H. Chan School of Public Health published a joint report called *The Global Economic Burden of Non-communicable Diseases.*[6] In this report, it was predicted that the cost of treating NCDs could bankrupt global healthcare systems by the year 2030. The preface of the joint report begins:

"Over the next 20 years, NCDs will cost more than US$30 trillion, representing 48% of global GDP in 2010, and pushing millions of people below the poverty line. Mental health conditions alone will account for the loss of an additional US$16.1 trillion over this time span, with dramatic impact on productivity and quality of life."

This means the cost of noncommunicable diseases between 2010 and 2030 would reach $47 trillion USD.

Professor Stefan Elbe puts this figure into perspective: "If you look at the United Kingdom, for example, its entire economic output in a single year is about 2.5 trillion US dollars. We are looking here at a cost that would be around twenty times that. The combined economic output of most countries in the world any given year is less than 1 trillion US dollars, and they are estimating a cost of US$47 trillion. It is a staggering sum indeed."

Commenting on the report, Olivier Raynaud, senior director of health at the World Economic Forum, warns, "The numbers indicate that noncommunicable diseases have the potential to not only bankrupt health systems but to also put a brake on the global economy."[7]

This economic prediction about the potential for the bankruptcy of our global health systems by the year 2030 is a worst-case scenario. However, if the worst-case scenario became our reality, how would our health services cope with the incredible rise in NCD patients? Would there be provision to treat everyone exactly when they needed treatment, or would some have to wait—or even not be treated at all?

If our workforce becomes sick, or burdened by caring for relatives who are sick, what further impact does that have on our economy? Fewer people working means fewer tax payments at the same time as increased pressure on welfare. The ripple effect could be never-ending.

The nature of NCDs—slow to develop, slow to diagnose, and often slow to treat—means that that it could be years before the general public really sits up and takes notice. Of course, by then it might be too late to change course. Nonetheless, their impact is so far-reaching and significant that experts now consider NCDs as much of a threat to populations as global pandemics. According to Professor Elbe, "We don't actually know when the next pandemic will occur, and how serious it will be. But we already have very good data that rates of noncommunicable diseases are rising at a fairly rapid rate. We know in the years ahead, they will have a pretty serious magnitude."

Scary words. The prevalence of NCDs is an urgent problem that demands urgent attention from our governments and global health-care systems. As Professor Elbe warns, "If we can't come up with cheap or cost-effective ways to address noncommunicable disease, we are going to be landed with quite a big socioeconomic problem."

What about infectious diseases?

So far we've explored how an altered infant microbiome or epigenetic changes at childbirth could be linked to an increased risk of NCDs. What about infectious diseases? According to Dr. Martin Blaser's hypothesis in *Missing Microbes*, loss of microbial diversity, combined with increased global interconnectivity (that is, our ability to travel freely and easily all over the world), also makes humans increasingly vulnerable to infectious disease and pandemics.

140

In his interview, Dr. Blaser said, "The risk is that, if our microbes get too depleted, then we will become quite susceptible to an invader, a pathogen that arises somewhere and that could spread through our communities all over the world and cause a terrible plague or pandemic. Pandemics have occurred in the past, and we're vulnerable to pandemics. There are many more of us; we live closer together. Because of jet planes and other forms of travel, we're no more than two days away from any village in the world. So we have a very large, interconnected community and that makes us more susceptible to pathogens. And now if our coastal defenses are weakening, if we don't have the diversity, then we're more susceptible."

Lee Jong-Wook, former director-general of WHO, said that infectious diseases "do not respect national boundaries."[8] In a UN magazine article about "National Security and Pandemics," the author Sara E. Davies writes that because infectious diseases have the ability to travel with us, across borders and boundaries, "they have the potential to weaken many societies, political systems, and economies simultaneously."[9]

Dr. Blaser called this scenario "antibiotic winter," a scenario that is very bleak: "It could really cause untold numbers of deaths. Now if this happened tomorrow, it's too late; there's not that much we can do about it. But if it were to happen 50 years from now, then we could find ways to protect against it and to restore our diversity. And that's what we have to do."

Should we be concerned for the future of humankind?

If the worst-case scenarios for NCDs and infectious diseases are not bad enough, there's also the possibility of an even deeper threat regarding our existence as a species. Remember that, the more interventions we have during pregnancy and birth, the greater the potential effects on our epigenetic heritage—every intervention could have ramifications for the health of our children, and of our children's children, and beyond (see chapter 6).

Professor Sue Carter, an internationally recognized expert in behavioral neuroendocrinology, expresses deep concern about not knowing the long-term effects of common interventions in childbirth: "I think we're in the midst of the largest experiment in human history, and we are not running the controlled studies that we need to know what the consequences are. I think that's unethical. I don't think it's acceptable. I'm not saying we shouldn't use the procedures and the medical advances that we have. I'm saying we owe it to the children that we're producing to know everything we can about what the consequences for that child of those procedures might be."

If we don't know the long-term effects of common interventions in childbirth for our children now, we certainly don't know what the effects may be many generations into the future, as pointed out by Professor Aleeca Bell: "It's irresponsible to be biologically manipulating this emerging system in the baby, this sensitive transitional system in the mother, and not know the consequences. We don't have a clue as to what kind of long-term, inheritable, intergenerational changes can take place."

According to Professor Carter, "My concerns for the future are that we're going to look back on this time period and say 'I can't believe we did that to our children.' We didn't know what we were doing."

Lesley Page expresses fears for what we might be doing in the future to the bonds within families: "We could be affecting the future of human society in ways we can't imagine at the moment. We might be affecting the relationship between the mother and the baby, the father and the baby, and the family relationships. And those in themselves will have a profound impact on society."

Could this be global warming for our species?

In June 1988, NASA scientist Dr. James E Hansen stood before the US Congress and testified that global warming had begun. Dr.

Hansen said that he was 99 percent certain that global warming "was not a natural variation but was caused by a build-up of carbon dioxide and other artificial gases in the atmosphere," according to a *New York Times* report of the event.[10]

In January 2015, the US Senate voted to accept that global warming was definitely happening, but the Senate stopped short of admitting that human activity was to blame.[11] This means that over a quarter of a century passed between Dr. Hansen's landmark testimony to the congressional committee and the US Senate finally voting to acknowledge the existence of global warming. This delay might have proven very costly in terms of potential environmental damage in the intervening 27 years.

Experts are now starting to draw parallels between what is happening on a macro level (the level of our planet with climate change) and what is happening on a micro level (the level of our microbiome with the loss of bacterial diversity).

According to Dr. Dominguez-Bello, "It's not just global warming at the level of the planet, but it's also in our bodies. We have impacted our macro ecosystems (our environmental ecosystems) but also we have impacted our micro ecosystems (our microbial ecosystems)." Professor Hannah Dahlen also uses the analogy of global warming: "Today, urgently, a bit like climate change, we might not have all the answers yet but we recognize this is a big problem." In the next chapter, we'll explore how we could stop a big problem today becoming an even bigger problem tomorrow.

Here's a summary of the main points we've covered in this chapter:

1. Noncommunicable disease (NCD) is currently the world's leading killer.

2. The World Health Organization has identified four main risk factors for NCD: smoking, misuse of alcohol, poor diet, and lack of exercise.

3. In 2011, a landmark joint report by the World Economic Forum and the Harvard T. H. Chan School of Public Health predicted the cost of NCD will be US $47 trillion by 2030, potentially bankrupting our global healthcare systems.

4. Children born by Cesarean section are at a significantly increased risk of developing multiple NCDs later in life. Cesarean section could be a factor accelerating us toward the disaster scenario outlined in the 2011 report.

5. Mode of birth is not currently listed as a contributing risk factor for NCD, but maybe it's time that it is.

6. Misuse of antibiotics, increased use of antibacterial products, plus rising rates of Cesarean section could all be contributing to key microbe species being lost, making populations more susceptible to pandemics. A worst-case scenario for the future is the antibiotic winter described by Dr. Martin Blaser.

 WORST-CASE SCENARIO

— NINE —

What Are
the Solutions?

The latest scientific research indicates that there's a connection between what happens in childbirth and a child's health in later life. If we interfere with vaginal birth (such as by using synthetic oxytocin to speed up labor) or we bypass it completely (as we do with elective Cesarean section), we could affect the microscopic events happening during childbirth. In all cases, we need more research to properly assess the impact on the baby's microbiome.

Children born by Cesarean section seem to be at a greater risk of developing certain immune-related conditions later in life. There are two plausible explanations for this: first, that a Cesarean section alters the optimal seeding of the baby's gut microbiome, and second, that a Cesarean section may cause certain epigenetic changes that are linked to negative health outcomes down the line. The long-term health risks associated with Cesarean sections could be down to one or the other of these possibilities, or it could be a combination of the two—or it could be something else altogether.

What is known for sure now is that the current level of noncommunicable disease (NCD) is a serious global problem. Populations have seen increases in rates of asthma, allergies, obesity, some cancers, cardiovascular disease, respiratory diseases, autoimmune conditions, and a variety of mental health conditions. The cost of treating these conditions is predicted to bankrupt global healthcare by 2030.

What's more, populations living in industrialized nations could already have lost a third of their bacterial diversity. According to Dr. Martin Blaser's hypothesis in *Missing Microbes*, this lack of bacterial diversity could be contributing to the rise of NCD and could also make us all much more vulnerable to infectious disease, with the result that, as nations, we are more susceptible to pandemics.

The rise in disease appears to relate largely to our diet and lifestyle. Crucially, though, it also appears to relate to the way we give birth. We can all do something about our health now by changing our diet and lifestyle. In addition, if expectant and future parents can become aware of and make a plan for the microscopic events that happen during birth, they have a chance to potentially change the health of future generations. Obviously, in all situations, the health of the baby and the mother is paramount, and for some parents having a Cesarean section is not a choice but a lifesaving necessity. Perhaps in the future, if a baby needs to be born by Cesarean section, it might become a supported option (provided the mother's screened vaginal microbiome tests negative for pathogens) for Dr. Maria Gloria Dominguez-Bello's swab-seeding technique to be employed to introduce the mother's vaginal microbes to the baby's microbiome. There might be other options available in the future to introduce the mother's intestinal microbes to the baby so they can join in the "colonization party" in the baby's gut.

As well as changing diet and lifestyle and maternity practices, there is a desperate need for more scientific research to provide more definitive answers about the long-term effects of common interventions in childbirth. When we interviewed Professor Aleeca Bell, she summed up the urgency and the enormity of the situation with one epic question, "If we don't take the bigger perspective and fund research that asks the smart questions, the big questions, the developmental questions, intergenerational questions, epigenetic questions—how do we know that we're not altering the course of humanity?"

What can we do to change things?

There are so many ways in which we can each, as individuals and as a community coming together, provide a catalyst for change and be part of the solution. Here is our own top-five "to-do" list—the things we want to do in the coming years to help bring about change.

1. Promote the key message that vaginal birth (where possible), immediate skin-to-skin contact, and breastfeeding for as long as possible are the best ways to ensure optimal seeding and feeding of the baby's gut microbiome.

   According to Dr. Rodney Dietert, vaginal birth, immediate skin-to-skin contact, and breastfeeding "should be on every birth plan. Every health professional, every medical professional tending to pregnant women should be aware of these things. They should be aware of the importance of the microbiome and supporting a healthy microbiome because that is what is going to influence health across a lifespan."

2. For babies that need to be born by Cesarean section, promote "woman-centered" or "natural" techniques (see "What can you do if you need a Cesarean section?" on page 89), including the baby being delivered straight onto the mother's chest, and full support for breastfeeding. Keep up to date with the research and viable solutions for restoring, or at least partially restoring, the microbiome of Cesarean section babies.

3. Champion more research into the long-term impact of medical interventions in childbirth.

   In order to persuade clinical practitioners, obstetricians, policy advisers, hospital managers, and perhaps the mainstream media that this is a subject that needs urgent global attention, we need a body of robust scientific research. We need strong evidence from high-quality research studies. As Dr. Blaser says, "We need to do more science to pinpoint the problems and also to develop the solutions."

So what's it going to take to have more research? There are many incremental steps to large-scale research, from looking at the published world literature for associations, through setting up and researching animal models, to setting up human research studies. All of this takes a lot of time. According to Dr. Neena Modi, "Studies could take 10, 20, 30, 40, 50, 60 years."

It will also take a serious amount of money. It could take large-scale investment in a comprehensive raft of medical research. As outlined by Dr. Blaser, "We have to invest in our health infrastructure in fundamental new ways. Just as we invest in building roads and bridges, these are projects too big for an individual person. We can pay for it now in research and higher cost for certain drugs, or we can pay for it later in the clinic. It's better to pay now."

And it will also require people to demand that the research gets under way. According to Dr. Modi, "I want to see parents demanding that this research is done. I want to see researchers developing the studies and pushing them out there. I want to see funders saying, 'We will fund long-term research, not just research that's got short-term outcomes.' I want to see policy makers saying, 'We can't write policy unless we have these studies.' We have a great opportunity before us now. Surely, surely together we should be seizing this?"

4. **Call for healthcare professionals to receive more training in the critical importance of the seeding and feeding of the baby's gut microbiome and the potential epigenetic changes happening during childbirth.**

Much of the information about the microscopic events happening during childbirth is so new that relatively few healthcare providers are aware of it, and even when they are aware of it, working it into hospital policy can in itself take years.

The process goes a bit like this: Researchers publish new research; the media or prominent people within the field champion it; and the organizations that oversee the maternity-related professions agree with and sanction it. Then, it gets worked into educating the healthcare professionals.

Sanctioned research might be incorporated into an academic curriculum for those students beginning their medical careers (in whatever form they might choose) but also, importantly, into the ongoing professional development of doctors, midwives, maternity nurses, lactation consultants, breastfeeding counselors, antenatal teachers, and other maternity professionals. The people already doing the jobs need to know, as well as those beginning on the path.

According to Dr. Philip Steer, "We're making steps forward in knowledge, both about the problems and about potentially how to deal with them, and of course the dilemma then is how do we get that information across to the professionals? And that really just has to come down to training, training, and training."

Once practitioners know about the research, it is absorbed into hospital policy. The whole process—from publishing the research to changing hospital policy—can take several years. That's why we need to start making noise about it straightaway.

5. **Raise awareness of all the information in this book within the wider public.**

We believe change happens in two ways. First, it happens from the top down—where political decision-makers (whether that is a politician on a national level or hospital managers on a local level) decide to change policies or make new laws according to their own agenda. Then everyone "below" adheres to those new laws. We can all help create change at the top by lobbying the political decision-makers, writing letters, making phone calls, starting an e-mail campaign, opening up an online petition, or even making an appointment to meet personally with those in a position of power.

Second, it happens from the bottom up. This is where grassroots activism can spread awareness using social media, or through special events, such as movie screenings of *Microbirth*, or through simple word of mouth.

When more people start hearing about the problem, more people are brought into the discussion. Then suddenly everyone is talking about it. It becomes a zeitgeist "thing." A change-maker at the top hears about the "thing" and decides they want to help find a solution. If they persuade their change-maker friends to join in with the solution, suddenly you hit the tipping point. A solution comes together. Someone—a government, an organization—finds the money to fund large-scale research programs. Maybe they find enough money to fund a whole raft of medical research. That's when change happens—when the will, money, and moment for change coincide.

Before us lies the possibility of change. If enough people ask enough questions about the long-term effects of modern childbirth practices, we can create enough weight to reach the tipping point, and then change will happen quickly. A distant possibility becomes a strong probability of change.

Remember, the whole process can start with something really simple, such as ordinary people asking questions. If you're pregnant, ask your healthcare provider (whether that's a midwife or a doctor) what systems are in place in your local area to help you seed and feed your baby's microbiome. If they don't have an answer or don't understand your question, by simply asking you will have motivated them to find an answer. If enough pregnant women ask the same question, eventually maternity managers will get to hear of it, and eventually systems will be put in place to ensure that every hospital, every birthing unit, and every midwife facilitates seeding and feeding at every birth.

What can we do in later life to minimize the possible effects of having been born by Cesarean section?

Dr. Dominguez-Bello's own daughter was born by C-section more than 20 years ago: "What I recommend to my daughter is to eat naturally; to avoid processed foods and in general to have a healthy lifestyle. If she has a healthy lifestyle, eats natural foods, and avoids antibiotics, she will one day, I hope, be a mother, and she needs to preserve the diversity that she has now in order to pass it to the next generation."

How can you get involved?

If you want to take an active part in this catalyst for change, what can you do? Ask questions. Seek answers. Look up the research and references mentioned in this book. Talk to friends, family, and colleagues about what you've learned. Talk to your local politician, clinician, midwife, and birth educator. Talk to your local parent group. Put on a *Microbirth* film screening. Get active on Twitter, Facebook, and other social media spreading the word. Start raising awareness now.

Dr. Blaser says this is an urgent public health crisis that demands action now: "We have to alert the public about what's going on so that they understand what the stakes are and where this is coming from."

The possibility of change lies before us. We have a choice. We can either sit back and do nothing, or we can take action now. The ramifications of this choice are beyond huge. They are epic.

According to Dr. Modi, "If how we're born affects our chances of long-term health, we need to do something about this now. We need to seize this opportunity to actually turn the course of human health and human history." Similarly, Professor Stefan Elbe sums up: "I think this is one where we are all going to have to pitch in together for the benefit of humanity."

Are we too late for change to happen? Dr. Jacquelyn Taylor thinks not: "I don't think we're ever too late. I think that we just need to move more rapidly." Dr. Dominguez-Bello agrees, "I don't think we are too late. I think we should work much more on these big issues, and we need a lot of support from a lot of scientists everywhere in the world. Because it's a global problem. It's not an American problem, it's not a European problem, it's not a Japanese problem, it's not a Chinese problem. It's a humanity problem."

What thoughts do our experts leave us with?

We always end our interviews by asking a couple of more reflective questions, about the person's hopes and fears for the future and about why they care about the subject. We were genuinely really taken aback by the fire, passion, and commitment to change within everyone's responses.

Dr. Martin Blaser: "I have both hopes and fears for the future. I'm afraid of scenarios like antibiotic winter; I'm afraid of the rises in diseases like obesity and diabetes and asthma. These are really alarming. But we use science to try to solve problems and to reach new solutions, and there's a lot of interest in the microbiome. I'm hoping that will lead to new insights and new solutions."

Dr. Jacquelyn Taylor: "I care because I'm a mother. I have children. I want my children to benefit from this type of research."

Dr. Rodney Dietert: "I care because we can't have 80 years with one chronic disease followed by another chronic disease followed by a third with more drugs, with more medical interventions. We don't have to have that. We can have prevention. We can make a difference. And we can do it in early life. We know what to do, and we should do it. It's so simple. The solution starts with birth."

Professor Aleeca Bell: "There's a pebble effect to birth in that women can feel they've given their baby the best possible start in life. They're alert, they're activated, they're engaged, they're empowered.

They give birth to babies that are alert. I truly believe that these evolutionary, conserved, physiological processes mean something during birth because they jump start systems in the mum and baby."

Professor Hannah Dahlen: "We haven't evolved to be the humans we are today without there being a significant benefit in the birth process. So I think we should begin with that. What is our purpose, and how were we meant to come into this world?"

Lesley Page: "We are not just concerned with a live healthy baby at birth, but we are also now beginning to see that we can have an impact on the future health of that baby, and an impact on the future of society."

Professor Anita Kozyrskyj: "In my previous career as a pharmacist in the neonatal intensive care unit, obviously I had a lot to do with the dispensing of medications to tiny babies. I am of the opinion that medications, although they are approved for their safety, are used in infants without proper consideration of their long-term effects. So one of the drugs I'm interested in is antibiotics. That's why I think my contribution will be important and why I care about research into whether the use of interventions such as antibiotic treatment causes changes to the footprint of the infant gut microbiota that may increase the risk of future disease."

Dr. Dominguez-Bello: "Our Westernized lifestyle and our technology has given us so many good things, but it also has collateral damage. We have to understand what is the impact and try to correct it before it's too late. My hope for the future is that we can restore the damage that we have already done. That the future for humanity will be bright and healthy."

What are our own hopes and fears?

If the tables were turned on us and we were asked the same final questions, about our hopes and fears for the future and about why we care, we would say . . .

Our greatest fear is that the world ignores our message. That nothing changes.

Our greatest hope is that, with this book, we sow a seed of possibility. That change can and does start to happen. At some point that possibility tips over into probability.

We've spent 3 years on this project. In some ways, it feels as if we are still at the beginning. If the seed of change takes hold, if this book can help galvanize a grassroots movement, then it could be a powerful catalyst for change. This catalyst can filter upward so that maternity-policy decision-makers and government lawmakers become interested in the microscopic events happening during childbirth. If that happens, then we will stand proud.

Obviously, we're not the only people asking questions about the microbiome and epigenetics. Along with the professors we have interviewed, many other academics, researchers, journalists, and authors are also exploring this rich and fast-moving area. With each discovery, with each academic paper, with each new film, with each new book publication, awareness of this urgent and important issue spreads ever wider.

We care that books like this one get written because we want them to be catalysts for change. More specifically, we want to make the world a better place for our daughter and for everyone's children. We hope that now that you've read our book and heard our stories and reasoning, you'll also want to do whatever you can to make the world a better place for everyone's children. We hope you'll stand with us as we change the world together.

Here's a summary of the main points we've covered in this chapter:

1. It's not too late to turn things around. But we have to act now.
2. We need to raise awareness that the optimal way to seed and feed a baby's microbiome is through vaginal

birth where possible, with immediate skin-to-skin contact and breastfeeding, again where possible.

3. If a baby needs to be born by Cesarean section, maternity systems could help the optimal seeding and feeding of a baby's microbiome by supporting immediate skin-to-skin contact with the mother in the operating theater, plus one-to-one support for exclusive breastfeeding. In addition, depending on Dr. Maria Gloria Dominguez-Bello's ongoing research, the swab-seeding technique for Cesarean section babies could be investigated as a possible supported option in the future (provided an expectant mother is screened to ensure she has a healthy vaginal microbiome).

4. If breastfeeding isn't a possibility, perhaps formula-milk manufacturers could be encouraged to develop products that better emulate the microbial (prebiotic and probiotic) and immune components of breast milk.

5. We all need to look again at our current antimicrobial practices, diets, and lifestyles—from the high use of antibiotics in farming and medicine to overuse of antibacterial products, and also, where necessary, to find ways to reduce the Cesarean section rate.

6. We need more fully funded and immediate research into the possible long-term effects of medical interventions in childbirth.

7. We need all current (and future) healthcare professionals and support services involved with birth to be made aware of, and trained in, the best practices to optimally seed and feed a baby's microbiome.

8. Change can happen from the top down by raising awareness of this issue with national and regional health policy makers and hospital managers.
9. Or change can happen from the bottom up, in which an expectant mother simply asks her health-care provider to help her seed and feed her baby's microbiome.
10. And then there's other direct action through social-media campaigns, petitions, and real-life events to bring more people into the conversation, to widen the discussion, and to hopefully come up with workable solutions. That's when a tipping point happens. That's when, together, we change the world.

WHAT'S NEXT FOR THE AUTHORS?

Acknowledgments

We are very grateful to the twelve professors featured in this book who agreed to be interviewed by us. We feel deep gratitude to them for kindly giving us permission for excerpts from their interview transcripts to be quoted in this book: Martin Blaser, Rodney Dietert, Maria Gloria Dominguez-Bello, Hannah Dahlen, Lesley Page, Aleeca Bell, Sue Carter, Anita Kozyrskyj, Neena Modi, Philip Steer, Jacquelyn Taylor, and Stefan Elbe.

We are also very grateful to Matthew Hyde, Cathy Warwick, and Julie Gerland for allowing us to interview them for *Microbirth*.

We are very grateful and appreciative of all those who contributed to our Indiegogo crowdfunding campaign to help fund the making of *Microbirth*.

Our deep appreciation to Martin Wagner, Zoë Blanc, and the rest of the brilliant publishing team at Pinter & Martin. And very special thanks to our Pinter & Martin book editor, Judy Barratt.

And lastly, our wondrous thanks to the equally brilliant publishing team at Chelsea Green Publishing including Margo Baldwin, Brianne Goodspeed, Pati Stone, and Angela Boyle.

Further Reading and Multimedia

The following books and articles have been suggested as further reading by Dr. Rodney Dietert, professor of immuntoxicology, Cornell University; author of *The Human Superorganism*. Many of the web articles are available in their complete form as free downloads.

BOOKS

Blaser, Martin J. *Missing Microbes: How the Overuse of Antibiotics Is Fueling Our Modern Plagues*. New York: Henry Holt and Co., 2014.

Dietert, Rodney R. *The Human Superorganism: How the Microbiome Is Revolutionizing the Pursuit of a Healthy Life*. New York: Dutton, 2016.

Dietert, Rodney R., and Janice Dietert. *Strategies for Protecting Your Child's Immune System: Tools for Parents and Parents-to-Be*. Singapore: World Scientific Publishing Company, 2010.

JOURNALS AND WEB ARTICLES

Almgren, M., T. Schlinzig, D. Gomez-Cabrero, A. Gunnar, M. Sundin, S. Johansson, M. Norman, and T. J. Ekström. "Cesarean Delivery and Hematopoietic Stem Cell Epigenetics in the Newborn Infant: Implications for Future Health?" *American Journal of Obstetrics and Gynecology* 21, no. 5 (2014): 502.e1-508.e80, doi:10.1016/j.ajog .2014.05.014.

Azad, Megan B., and Anita L. Kozyrskyj. "Perinatal Programming of Asthma: The Role of Gut Microbiota." *Clinical and Developmental Immunology* (2012): 932072, doi:10.1155/2012/932072.

Azad, M.B., T. Konya, H. Maughan, D. S. Guttman, C. J. Field, R. S. Chari, M. R. Sears, A. B. Becker, J. A. Scott, A. L. Kozyrskyj, and CHILD Study investigators. "Gut Microbiota of Healthy

Canadian Infants: Profiles by Mode of Delivery and Infant Diet at 4 Months." *Canadian Medical Association Journal* 185, no. 5 (2013): 385–94, doi:10.1503/cmaj.121189.

Blaser, Martin J. "Who Are We? Indigenous Microbes and the Ecology of Human Diseases." *The European Molecular Biology Organization Reports* 7, no. 10 (2006): 956–60, doi:10.1038/sj .embor.7400812.

Bloom, D. E., E. T. Cafiero, E. Jané-Llopis, S. Abrahams-Gessel, L. R. Bloom, S. Fathima, A. B. Feigl, T. Gaziano, M. Mowafi, A. Pandya, K. Prettner, L. Rosenberg, B. Seligman, A. Z. Stein, and C. Weinstein. "The Global Economic Burden of Noncommunicable Diseases." *Geneva: World Economic Forum* (2013).

Borre, Y. E., R. D. Moloney, G. Clarke, T. G. Dinan, and J. F. Cryan. "The Impact of Microbiota on Brain and Behavior: Mechanisms & Therapeutic Potential." *Advances in Experimental Medicine and Biology* 817 (2014): 373–403, doi:10.1007/978-1-4939-0897 -4_17.

Decker, E., M. Hornef, and S. Stockinger. "Cesarean Delivery Is Associated with Celiac Disease but Not Inflammatory Bowel Disease in Children." *Gut Microbes* 2, no. 2 (2011): 91–98, doi:10.4161/gmic.2.2.15414.

Dietert, Rodney R. "Developmental Immunotoxicity, Perinatal Programming, and Noncommunicable Diseases: Focus on Human Studies." *Advances in Medicine* (2014), doi: 10.1155/2014/867805.

Dietert, Rodney R. "Early Immune Education." *Eureka Science* (April 22, 2014): http://www.criver.com/about-us/eureka/blog/april -2014/early-immune-education.

Dietert, Rodney R. "Natural Childbirth and Breastfeeding as Preventive Measures of Immune-Microbiome Dysbiosis and Misregulated Inflammation." *Journal of Ancient Diseases & Preventive Remedies* 1, no. 103 (2013), doi:10.4172/2329-8731.1000103.

Dietert, Rodney R., and Janice M. Dietert. "The Completed Self: An Immunological View of the Human-Microbiome Superorganism and Risk of Chronic Diseases." *Entropy* 14 (2012): 2036–65, doi:10.3390/e14112036.

Dietert, Rodney R., J. C. DeWitt, D. R. Germolec, and J. T. Zelikoff. "Breaking Patterns of Environmentally Influenced Disease for

Health Risk Reduction: Immune Perspectives." *Environmental Health Perspectives* 118, no. 8 (2010): 1091–99, doi:10.1289/ehp .1001971.

Dominguez-Bello, Maria Gloria, E. K. Costello, M. Contreras, M. Magris, G. Hidalgo, N. Fierer, and R. Knight. "Delivery Mode Shapes the Acquisition and Structure of the Initial Microbiota Across Multiple Body Habitats in Newborns." *Proceedings of the National Academy of Sciences U.S.A.* 107, no. 26 (2010): 11971–75.

Guinane, C. M., and P. D. Cotter. "Role of the Gut Microbiota in Health and Chronic Gastrointestinal Disease: Understanding a Hidden Metabolic Organ." *Therapeutic Advances in Gastroenterology* 6, no. 4 (2013): 295–308, doi:10.1177/1756283X13482996.

Huang, L., Q. Chen, Y. Zhao, W. Wang, F. Fang, and Y. Bao. "Is Elective Cesarean Section Associated with a Higher Risk of Asthma? A Meta-Analysis." *Journal of Asthma* (August 27, 2014): 1–10, doi:10.3 109/02770903.2014.952435.

Human Microbiome Project Consortium. "Structure, Function and Diversity of the Healthy Human Microbiome." *Nature* 486, no. 7402 (2012): 207–14, doi:10.1038/nature11234.

Jakobsson, H. E., T. R. Abrahamsson, M. C. Jenmalm, K. Harris, C. Quince, C. Jernberg, B. Björkstén, L. Engstrand, and A. F. Andersson. "Decreased Gut Microbiota Diversity, Delayed Bacteroidetes Colonisation and Reduced Th1 Responses in Infants Delivered by Caesarean Section." *Gut* 63, no. 4 (2014): 559–66, doi:10.1136/gutjnl-2012-303249.

Mårild, K., O. Stephansson, S. Montgomery, J. A. Murray, and J. F. Ludvigsson. "Pregnancy Outcome and Risk of Celiac Disease in Offspring: A Nationwide Case-Control Study." *Gastroenterology* 142, no. 1 (2012): 39–45; e3, doi:10.1053/j.gastro.2011.09.047.

Markle, J. G., D. N. Frank, K. Adeli, M. von Bergen, and J. S. Danska. "Microbiome Manipulation Modifies Sex-Specific Risk for Autoimmunity." *Gut Microbes* 5, no. 4 (2014): 485–93, doi:10.4161 /gmic.29795.

Mejía-León, M. E., J. F. Petrosino, N. J. Ajami, M. G. Domínguez-Bello, and A. M. de la Barca. "Fecal Microbiota Imbalance in Mexican Children with Type 1 Diabetes." *Scientific Reports* 4 (2014): 3814, doi:10.1038/srep03814.

Vajro, P., G. Paolella, and A. Fasano. "Microbiota and Gut-Liver Axis: Their Influences on Obesity and Obesity-Related Liver Disease." *Journal of Pediatric Gastroenterology and Nutrition* 56, no. 5 (2014): 461–8, doi:10.1097/MPG.0b013e318284abb5.

Walker, W. A. "Initial Intestinal Colonization in the Human Infant and Immune Homeostasis." *Annals of Nutrition and Metabolism* 63, no. 2 (2014): 8–15, doi:10.1159/000354907.

Weng, M., and W. A. Walker. "The Role of Gut Microbiota in Programming the Immune Phenotype." *The Journal of Developmental Origins of Health and Disease* 4, no. 3 (2013): 203–14.

URLs for the videos from *Microbirth*

Introduction: "Meet The Authors" vimeo.com/154871900

Chapter 1: "What Is the Microbiome?" vimeo.com/151882942

Chapter 2: "How the Microbiome Is Seeded at Birth" vimeo.com /151882943

Chapter 3: "Breastfeeding and the Microbiome" vimeo.com /151882944

Chapter 4: "Cesarean Section and the Microbiome" vimeo.com /151885515

Chapter 5: "Immune System and the Microbiome" vimeo.com /151885514

Chapter 6: "Epigenetics Explained" vimeo.com/151885517

Chapter 7: "Dr. Dietert's 'Tree of Disease'" vimeo.com/151885518

Chapter 8: "Worst-Case Scenario" vimeo.com/151885516

Chapter 9: "What's Next for the Authors?" vimeo.com/152602719

References

Provided by Professor Hannah Dahlen, professor of midwifery, Western Sydney University

Dahlen, Hannah Grace. "Why Being Born Is Good for You." *The Practising Midwife* 18, no. 4 (2015): 10–13.

———. "Can Love and Science Co-exist in This Debate?" *International Journal of Birth and Parent Education* 2, no. 2 (2015): 40–41.

Dahlen, Hannah Grace, Soo Downe, Holly Powell Kennedy, and Maralyn Foureur. "Is Society Being Reshaped on a Microbiological and Epigenetic Level by the Way Women Give Birth?" *Midwifery* 30, no. 12 (2014): 1149–51, doi:10.1016/j.midw.2014.07.007.

Dahlen, Hannah Grace, Soo Downe, M. L. Wright, Holly Powell Kennedy, and Jacquelyn Y. Taylor. "Childbirth and Consequent Atopic Disease: Emerging Evidence on Epigenetic Effects Based on the Hygiene and EPIIC Hypotheses." *BMC Pregnancy and Childbirth* 16, no. 1 (2015): 4, doi:10.1186/s12884-015-0768-9.

Provided by Jacquelyn Y. Taylor, associate professor, Robert Wood Johnson Foundation; nurse faculty scholar alumna, University of Yale

Anderson, C. M., J. Ralph, L. Johnson, A. Scheett, M. L. Wright, Jacquelyn Y. Taylor, J. E. Ohm, and E. Uthus. "First Trimester Vitamin D Status and Placental Epigenomics in Preeclampsia Among Northern Plains Primiparas." *Life Sciences* 129 (2015): 10–15, doi:10.1016/j.lfs.2014.07.012. PubMed PMID: 25050465.

Capitulo, K., V. Klein, and M. Wright. "Should Prophylactic Antibiotics Be Routinely Given to a Mother Before a Cesarean Birth?" *MCN American Journal of Maternal Child Nursing* 38, no. 5 (2013): 266–67, doi:10.1097/NMC.0b013e31829b3d68. PubMed PMID: 23958615.

Clark, A. E., M. Adamian, and Jacquelyn Y. Taylor. "An Overview of Epigenetics in Nursing." *Nursing Clinics of North America* 48 (2013): 649–59, doi:10.1016/j.cnur.2013.08.004. PubMed Central PMCID: PMC3873714.

Dahlen, Hannah Grace, Soo Downe, M. L. Wright, H. P. Kennedy, and Jacquelyn Y. Taylor. "Childbirth and Consequent Atopic Disease: Emerging Evidence on Epigenetic Effects Based on the Hygiene and EPIIC Hypotheses." *BioMed Research International* 16, no. 1 (2016): 4.

Dahlen, Hannah Grace, H. P. Kennedy, C. M. Anderson, A. F. Bell, A. Clark, Maralyn Foureur, J. E. Ohm, A. M. Shearman, Jacquelyn Y. Taylor, M. L. Wright, and Soo Downe. "The EPIIC Hypothesis: Intrapartum Effects on the Neonatal Epigenome and Consequent Health Outcomes." *Medical Hypotheses* 80, no. 5 (2013): 656–62, doi:10.1016/j.mehy.2013.01.017. PubMed PMID: 23414680.

Taylor, Jacquelyn Y., M. Wright, C. Crusto, and Y. V. Sun. "The Intergenerational Impact of Genetic and Psychological Factors on Blood Pressure Study (InterGEN): Design and Methods for Complex DNA Analysis." *Biological Research for Nursing* (2016), doi:10.1177/1099800416645399.

Wright, M. L., and A. R. Starkweather. "Antenatal Microbiome: Potential Contributor to Fetal Programming and Establishment of the Microbiome in Offspring." *Nursing Research* 64, no. 4 (2015): 306–19, doi:10.1097/NNR.0000000000000101. PubMed PMID: 26035769.

Provided by Anita Kozyrskyj, professor, Department of Pediatrics, University of Alberta

Azad, M. B., T. Konya, R. R. Persaud, D. S. Guttman, R. S. Chari, C. J. Field, M. R. Sears, P. J. Mandhane, S. E. Turvey, P. Subbarao, A. B. Becker, J. A. Scott, Anita L. Kozyrskyj, and the CHILD Study investigators. "Impact of Maternal Intrapartum Antibiotics, Method of Birth and Breastfeeding on Gut Microbiota During the First Year of Life: A Prospective Cohort Study." *BJOG: An International Journal of Obstetrics and Gynaecology* (September 28, 2015), doi:10.1111/1471-0528.13601.

REFERENCES

Azad, M. B., T. Konya, D. S. Guttman, C. J. Field, M. R. Sears, K. T. HayGlass, P. J. Mandhane, S. E. Turvey, P. Subbarao, A. B. Becker, J. A. Scott, Anita L. Kozyrskyj, and the CHILD Study investigators. "Infant Gut Microbiota and Food Sensitization: Associations in the First Year of Life." *Clinical and Experimental Allergy* 45 (2015): 632–43, doi:10.1111/cea.12487.

Bridgman, S. L., T. Konya, M. B. Azad, M. R. Sears, A. B. Becker, S. E. Turvey, P. J. Mandhane, P. Subbarao, and the CHILD Study investigators, J. A. Scott, C. J. Field, and A. L. Kozyrskyj. "Infant Gut Immunity: A Preliminary Study of IgA Associations with Breastfeeding." *Journal of Developmental Origins of Health and Disease* 7, no. 1 (2016): 103–7, doi:10.1017/S2040174415007862.

Koleva, P. T., S. L. Bridgman, and A. L. Kozyrskyj. "The Infant Gut Microbiome: Evidence for Obesity Risk and Dietary Intervention." *Nutrients* 7, no. 4 (April 2015): 2237–60, doi:10.3390/nu7042237.

Kozyrskyj, A. L., R. Kalu, P. T. Koleva, and S. L. Bridgman. "Fetal Programming of Overweight Through the Microbiome: Boys Are Disproportionately Affected." *Journal of Developmental Origins of Health and Disease* (June 2015): 1–10, doi:10.1017/S2040174415001269.

Mastromarino, P., D. Capobianco, A. Miccheli, G. Praticò, G. Campagna, N. Laforgia, T. Capursi, and M. E. Baldassarre. "Administration of a Multistrain Probiotic Product (VSL#3) to Women in the Perinatal Period Differentially Affects Breast Milk Beneficial Microbiota in Relation to Mode of Delivery." *Pharmacological Research* (2015): 95–96, 63–70, doi:10.1016/j.phrs.2015.03.013.

Provided by Professor Stefan Elbe, professor of international relations, University of Sussex; director of Centre for Global Health Policy

Elbe, Stefan. *Security and Global Health: Toward the Medicalization of Insecurity.* Australia: Polity Press, 2010.

Provided by Dr. Neena Modi, professor of neonatal medicine, Imperial College, London

Darmasseelane, K., M. J. Hyde, S. Santhakumaran, C. Gale, and N. Modi. "Mode of Delivery and Offspring Body Mass Index,

Overweight and Obesity in Adult Life: A Systematic Review and
Meta-analysis." *PLoS ONE* 9, no. 2 (2014): e87896, doi:10.1371
/journal.pone.0087896. PubMed PMID: 24586295.

Svensson, E., M. J. Hyde, N. Modi, and V. Ehrenstein. "Caesarean
Section and Body Mass Index Among Danish Adult Men." *Obesity*
21, no. 3 (Silver Spring, March 2013): 429–33, doi:10.1002/
oby.20310. PubMed PMID: 23408746.

**Provided by Professor Aleeca Bell, assistant professor,
Department of Women, Children, and Family Health Science,
University of Illinois at Chicago**

Bell, A. F., C. S. Carter, C. D. Steer, J. Golding, J. M. Davis, A. D.
Steffen, et al. "Interaction Between Oxytocin Receptor DNA
Methylation and Genotype Is Associated with Risk of Postpartum
Depression in Women Without Depression in Pregnancy." *Frontiers
in Genetics* 6 (2015): 243, doi:10.3389/fgene.2015.00243.

Bell, A. F., C. S. Carter, J. M. Davis, J. Golding, O. Adejumo, M. Pyra,
et al. "Childbirth and Symptoms of Postpartum Depression and
Anxiety: A Prospective Birth Cohort Study." *Archives of Women's
Mental Health* (July 23, 2015), doi:10.1007/s00737-015-0555-7.

Bell, A. F., E. N. Erickson, and C. S. Carter. "Beyond Labor: The
Role of Natural and Synthetic Oxytocin in the Transition to
Motherhood." *Journal of Midwifery & Women's Health* 59, no. 1
(2014): 35–42, doi:10.1111/jmwh.12101.

Notes

Introduction

1. Rodney R. Dietert and Janice M. Dietert, "The Completed Self: An Immunological View of the Human-Microbiome Superorganism and Risk of Chronic Diseases," *Entropy* 14, no. 11 (2012): 2036–65.
2. Dr. Rodney Dietert was awarded the James G. Wilson Publication Award for his paper published in *Birth Defects Research*: "The Microbiome in Early Life: Self-Completion and Microbiota Protection as Health Priorities." See Nicole Chavez, "Rodney Dietert, PhD, Receives James G. Wilson Publication Award," *Birth Defects Research Connection*, The Teratology Society (May 27, 2015): http://connection.teratology.org/p/bl /et/%20blogid=17&blogaid=440.
3. *Doula! The Film* (Alto Films Ltd., 2010), DVD: http://doulafilm .com.
4. *Freedom for Birth* (Alto Films Ltd., 2012): http://freedomforbirth .com. Released on September 20, 2012, with over 1,000 public screenings worldwide. It has been estimated that more than 100,000 people saw the film on day one.
5. *Microbirth* (Alto Films Ltd., 2014): http://microbirth.com.
6. "Martin Blaser—New York University," YouTube video, 2:23, part of a series of interviews given at the International Human Microbiome Congress in March 2012, organized by the MetaHIT project, posted by "Yohanan Winogradsky," May 15, 2012, https://youtu.be/VAn03UooZOA.
7. Raquel Maurier, "C-Section, Formula Feeding Affect Babies' Gut Bacteria," *University of Alberta* (news website), February 11, 2013, http://uofa.ualberta.ca/news-and-events/newsarticles /2013/february/csectionformulaaffectbabiesgutbacteria.
8. Karthik Darmasseelane, Matthew J. Hyde, Shalini Santhakumaran, Chris Gale, and Neena Modi, "Mode of

Delivery and Offspring Body Mass Index, Overweight and Obesity in Adult Life: A Systematic Review and Meta-Analysis," *PLoS ONE* 9, no. 2 (February 26, 2014): e87896, doi:10.1371/journal.pone.0087896.

Chapter 1: What Is the Human Microbiome?

1. Peter Andrey Smith, "Is Your Body Mostly Microbes? Actually, We Have No Idea," *Boston Globe*, September 14, 2014, http:// www.bostonglobe.com/ideas/2014/09/13/your-body-mostly -microbes-actually-have-idea/qlcoKot4wfUXecjeVaFKFN /story.html.

2. "NIH Human Microbiome Project Defines Normal Bacterial Makeup of the Body," *National Institutes of Health*, June 13, 2012, http://www.nih.gov/news/health/jun2012/nhgri -13.htm.

3. Robert Bowers, Amy Sullivan, Elizabeth Costello, Jeff Collett Jr., Rob Knight, and Noah Fierer, "Sources of Bacteria in Outdoor Air Across Cities in the Midwestern United States," *Applied and Environmental Microbiology* 77, no. 18 (2011): 6350–56, doi:10.1128/AEM.05498-11.

4. Debbie A. Lewis, Richard Brown, Jon Williams, Paul White, S. Kim Jackobson, Julian R. Marchesi, and Marcus J. Drake, "The Human Urinary Microbiome; Bacterial DNA in Voided Urine of Asymptomatic Adults," *Frontiers in Cellular and Infection Microbiology* 3, no. 41 (August 15, 2013), doi:10.3389/fcimb .2013.00041.

5. William G. Branton, Kritofor K. Ellestad, Ferdinand Maingat, B. Matt Wheatley, Erling Rud, René L. Warren, Robert A. Holt, Michael G. Surette, and Christopher Power, "Brain Microbial Populations in HIV/AIDS: α-Proteobacteria Predominate Independent of Host Immune Status," *PLoS ONE* 8, no. 1 (January 23, 2013): e54673, doi:10.1371/journal .pone.0054673.

6. F. Kong and R. P. Singh, "Disintegration of Solid Foods in Human Stomach," *Journal of Food Science* 73, no. 5 (June 7, 2008): R67–80, doi:10.1111/j.1750-3841.2008.00766.x.

7. Jaladanki N. Rao and Jian-Ying Wang, "Intestinal Architecture and Development," *Regulation of Gastrointestinal Mucosal Growth* (San Rafael, California: Morgan & Claypool Life Sciences, 2010): http://www.ncbi.nlm.nih.gov/books/NBK54098.
8. Alison Stephen and J. H. Cummings, "The Microbial Contribution to Human Faecal Mass," *Journal of Medical Microbiology* 13, no. 1 (February 1980): 45–56, doi:10.1099/00222615-13-1-45.
9. "Food Allergy Basics: Facts and Statistics," *FARE—Food Allergy Research and Education*, http://www.foodallergy.org/facts -and-stats.
10. In the study, type 1 diabetes had risen from 1.5 cases per 1,000 children in 2002 to 2.3 cases per 1,000 in 2013. "More US Kids Had Type 1 Diabetes but Researchers Don't Know Why," *MedlinePlus* (December 17, 2015): https://web.archive.org /web/20160104205615, https://www.nlm.nih.gov/medlineplus /news/fullstory_156281.html.
11. "What is Celiac Disease?," *Celiac Disease Foundation*, http:// celiac.org/celiac-disease/what-is-celiac-disease.
12. "Overweight and Obesity Statistics," *National Institute of Diabetes and Digestive and Kidney Diseases*, last modified October 2012, http://www.niddk.nih.gov/health-information/health -statistics/Pages/overweight-obesity-statistics.aspx.
13. "Heartburn and Gastro-Oesophageal Reflux Disease (GORD)," *NHS Choices*, last modified February 3, 2016, http://www.nhs. uk/conditions/Gastroesophageal-reflux-disease/Pages /Introduction.aspx.
14. Jose C. Clemente, Erica C. Pehrsson, Martin J. Blaser, Kuldip Sandhu, Zhan Gao, Bin Wang, Magda Magris, et al., "The Microbiome of Uncontacted Amerindians," *Science Advances* 1, no. 3 (April 17, 2015) ii: e1500183, doi:10.1126/sciadv.1500183.
15. L. Charles Bailey, Christopher B. Forrest, Peixin Zhang, Thomas M. Richards, Alice Livshits, and Patricia A. DeRusso, "Association of Antibiotics in Infancy with Early Childhood Obesity," *JAMA Pediatrics* 168 no. 11 (November 2014): 1063-69, doi:10.1001/jamapediatrics.2014.1539.

16. Thomas P. Van Boeckel, Charles Brower, Marius Gilbert, Bryan T. Grenfell, Simon A. Levin, Timothy P. Robinson, et al., "Global Trends in Antimicrobial Use in Food Animals," *Proceedings of the National Academy of Sciences of the USA* 112, no. 18 (November 21, 2014): 5649–54, doi:10.1073/pnas.1503141112.

17. Bazian, "Antibiotic Use in Farm Animals 'Threatens Human Health,'" *NHS Choices*, December 9, 2015, http://www.nhs.uk/news/2015/12December/Pages/Antibiotic-use-in-farm-animals-threatens-human-health.aspx.

18. Bailey et al., "Association of Antibiotics in Infancy with Early Childhood Obesity," 1063–69.

19. Van Boeckel et al., "Global Trends in Antimicrobial Use in Food Animals," 5649–54.

20. Helen Browning, "Overuse of Antiobiotics in Farming," *Soil Association*, December 22, 2015, https://www.soilassociation.org/blogs/2015/december/22/overuse-of-antibiotics-in-farming.

21. "Antiobiotics," *Microbiology Online*, http://www.microbiologyonline.org.uk/about-microbiology/microbes-and-the-human-body/antibiotics.

    Some people believe that penicillin was actually first discovered by a physician called John Tyndall in 1875; "A Brief History of Penicillin," *The Role of Chemistry in History*, March 26, 2008, http://itech.dickinson.edu/chemistry/?p=107.

22. "Discovery and Development of Penicillin," American Chemical Society International Historic Chemical Landmarks, last accessed September 6, 2016, http://www.acs.org/content/acs/en/education/whatischemistry/landmarks/flemingpenicillin.html.

23. "A Brief History of Penicillin," *The Role of Chemistry in History*.

24. Mike Turner, "Antibiotic Resistance: 6 Diseases That May Come Back to Haunt Us," *The Guardian*, May 9, 2014, http://www.theguardian.com/commentisfree/2014/may/09/6-diseases-becoming-resistant-to-antibiotics.

    "Antimicrobial Resistance: Global Report on Surveillance 2014," *World Health Organization*, April, 2014, http://www.who.int/drugresistance/documents/surveillancereport/en.

25. "Vaginal Thrush," *NHS Choices*, last updated April 3, 2016, http://www.nhs.uk/Conditions/Thrush/Pages/Causes .aspx.

Chapter 2: What Do Bacteria Have to Do with Birth?

1. Alice Park, "Babies in the Womb Aren't So Sterile After All," *TIME*, December 28, 2015, http://time.com/4159249/baby -microbiome-womb.
2. Molly J. Stout, Bridget Conlon, Michele Landeau, Iris Lee, Carolyn Bower, Zhao Qiuhong, et al., "Identification of Intracellular Bacteria in the Basal Plate of the Human Placenta in Term and Preterm Gestations," *American Journal of Obstetrics and Gynecology* 208, no. 3 (January 21, 2013): 226, e221–e227, doi:10.1016/j.ajog.2013.01.018.

   Kjersti Aagaard, Jun Ma, Kathleen M. Antony, Radhika Ganu, Joseph Petrosino, and James Versalovic, "The Placenta Harbors a Unique Microbiome," *Science Translational Medicine* 6, no. 237 (May 21, 2014): 237ra65, doi:10.1126/scitranslmed .3008599.
3. Bin Cao, Molly J. Stout, Iris Lee, and Indira U. Mysorekar, "Placental Microbiome and Its Role in Preterm Birth," *NeoReviews* 15, no. 12 (December 1, 2014): e537–e545, doi:10.1542/neo.15-12-e537.
4. Aagaard et al., "The Placenta Harbors a Unique Microbiome."
5. Park, "Babies in the Womb Aren't So Sterile After All."
6. Noel T. Mueller, Elizabeth Bakacs, Joan Combellick, Zoya Grigoryan, and Maria G. Dominguez-Bello, "The Infant Microbiome Development: Mom Matters," *Trends in Molecular Medicine* 21, no. 2 (January 8, 2015): 109–17, doi:10.1016/j.mol med.2014.12.002.
7. Ibid.
8. Ibid.
9. Omry Koren, Julia K. Goodrich, Tyler C. Cullender, Aymé Spor, Kirsi Laitinen, Helene Kling Bäckhed, et al., "Host Remodeling of the Gut Microbiome and Metabolic Changes During Pregnancy," *Cell* 150 (August 2, 2012): 470–80, doi:10.1016/j .cell.2012.07.008.

10. Mueller et al., "The Infant Microbiome Development: Mom Matters," 109–17.

11. Martin J. Blaser, *Missing Microbes: How the Overuse of Antibiotics Is Fueling Our Modern Plagues* (New York: Henry Holt and Co., 2014): 94.

12. Dietert, "The Completed Self: An Immunological View of the Human-Microbiome Superorganism and Risk of Chronic Diseases."

13. Keith M. Godfrey and David J. P. Barker, "Fetal Programming and Adult Health," *Public Health Nutrition* 4, no. 2b (April 2001, published online September 1, 2007): 611–24, doi:10.1079/PHN 2001145.

14. Daria A. Kashtanova, Anna S. Popenko, Olga N. Tkacheva, Alexander B. Tyakht, Dimitry G. Alexeev, and Sergey A. Boytsov, "Association Between the Gut Microbiota and Diet: Fetal Life, Early Childhood, and Further Life," *Nutrition* 32, no. 6 (published online December 31, 2015): 650-7, doi:10.1016/j .nut.2015.12.037.

15. Erika von Mutius, "99th Dahlem Conference on Infection, Inflammation and Chronic Inflammatory Disorders: Farm Lifestyles and the Hygiene Hypothesis," *Clinical and Experimental Immunology* 160, no. 1 (April 2010): 130–35, doi:10.1111/j.1365-2249.2010.04138.x.

16. "En Caul Baby Birth," *BabyMed*, http://www.babymed.com /en-caul-birth.

17. Blaser, *Missing Microbes*, 95.

18. Juan Miguel Rodriguez, Kiera Murphy, Catherine Stanton, R. Paul Ross, Olivia I. Kober, Nathalie Juge, et al., "The Composition of the Gut Microbiota Throughout Life, with an Emphasis on Early Life," *Microbial Ecology in Health and Disease* 26 (February 2, 2015): 26050, http://www.microbecolhealth dis.net/index.php/mehd/article/view/26050.

Research by Backhed et al. in 2015 suggests that the act of stopping breastfeeding (rather than simply introducing solids) could play a part in determining when the microbiome stabilizes, but how and why is still unclear.

19. Robert E. Ward, Milady Niñonuevo, David A. Mills, Carlito B. Lebrilla, J. Bruce German, "In Vitro Fermentation of Breast Milk Oligosaccharides by *Bifidobacterium infantis* and *Lactobacillus gasseri*," *Applied and Environmental Microbiology* 72, no. 6 (June 2006): 4497–99, doi:10.1128/AEM.02515-05.
20. Kashtanova et al., "Association Between the Gut Microbiota and Diet: Fetal Life, Early Childhood, and Further Life."
21. Noah Voreades, Anne Kozil, and Tiffany L. Weir, "Diet and the Development of the Human Intestinal Microbiome," *Frontiers in Microbiology* 5 (September 22, 2014): 494, doi:10.3389/fmicb .2014.00494.

Chapter 3: Breast Milk or Formula?

1. A. M. Widström, A. B. Ransjo-Arvidson, K. Christensson, A.-S. Matthiesen, J. Winberg, and K. Uvnäs-Moberg, "Gastric Suction in Healthy Newborn Infants: Effects on Circulation and Developing Feeding Behaviour," *Acta Paediatrica Scandinavica* 76 (July 1987): 566–72, doi:10.1111/j.1651-2227.1987.tb10522.x.
2. A. M. Widström, G. Lilja, P. Aaltomaa-Michalias, A. Dahllöf, M. Lintula, and E. Nissen, "Newborn Behaviour to Locate the Breast When Skin-to-Skin: A Possible Method for Enabling Early Self-Regulation," *Acta Paediatrica Scandinavica* 100, no. 1 (August 10, 2010): 79–85, doi:10.1111/j.1651-2227.2010.01983.x.
3. "Components of Colostrum," *La Belle Colostrum*, last accessed September 7, 2016, http://labelleinc.com/human-health /components-of-colostrum.
4. Blaser, *Missing Microbes*, 94.
5. Raul Cabrera-Rubio, M. Carmen Collado, Kirsi Laitinen, Seppo Salminen, Erika Isolauri, and Alex Mira, "The Human Milk Microbiome Changes Over Lactation and Is Shaped by Maternal Weight and Mode of Delivery," *American Journal of Clinical Nutrition* 96, no. 3 (July 25, 2012): 544–51, doi:10.3945 /ajcn.112.037382.
6. "Our Formula Milks," *HiPP Organic*, last accessed September 7, 2016, http://www.hipp.co.uk/products/our-baby-milks.

Chapter 4: What Is the Impact of Cesarean Section on the Microbiome?

1. "Achievements in Public Health, 1900–1999: Healthier Mothers and Babies," *Morbidity and Mortality Weekly Report (MMWR)*, US Centers for Disease Control and Prevention, October 1, 1999, http://www.cdc.gov/mmwr/preview/mmwrhtml/mm4838a2.htm.

2. Nicholas J. Kassebaum, Amelia Bertozzi-Villa, Megan S. Coggeshall, Katya A. Shackelford, Caitlyn Steiner, Kyle R. Heuton, et al., "Global, Regional, and National Levels and Causes of Maternal Mortality During 1990–2013: A Systematic Analysis for the Global Burden of Disease Study," *The Lancet* 384, no. 9947 (September 13, 2014): 980–1004, doi:10.1016/S0140-6736(14)60696-6.

3. Ibid.

4. "NHS Maternity Statistics—England, 2013–14," *Health and Social Care Information Centre*, January 28, 2015, http://www.hscic.gov.uk/catalogue/PUB16725.

5. "Casearean Sections Should Only Be Performed When Medically Necessary," World Health Organization news release, April 10, 2015, http://www.who.int/mediacentre/news/releases/2015/caesarean-sections/en.

6. "Caesarean Section," *OECD Data*, https://data.oecd.org/healthcare/caesarean-sections.htm.

7. "New Rules to Curb 'Epidemic' of Cesarean Births in Brazil (Update)," The Associated Press, *MedicalXpress*, January 7, 2015, http://medicalxpress.com/news/2015-01-curb-epidemic-cesarean-births-brazil.html.

8. Gunnar Holmgren, Lennart Sjöholm, and Michael Stark, "The Misgav Ladach Method for Cesarean Section: Method Description," *Acta Obstetrica et Gynecologica Scandinavica* 78, no. 7 (July 1999): 615–21, http://www.ncbi.nlm.nih.gov/pubmed/10422908, PubMed PMID: 10422908.

9. National Institute for Health and Care Excellence (NICE), "Guidance," in *NICE Guidelines [CG132]: Caesarean Section*, last updated August 2012, http://www.nice.org.uk/guidance/cg132/chapter/guidance.

10. Emma L. Barber, Lisbet Lundsberg, Kathleen Belanger, Christian M. Pettker, Edmond F. Funai, and Jessica L. Illuzzi, "Contributing Indications to the Rising Cesarean Delivery Rate," *Obstetrics and Gynecology* 118, no. 1 (July 2011): 29–38, doi:10.1097/AOG.0b013e31821e5f65.

11. Betty Kovacs, "Probiotics: What Are Prebiotics and Synbiotics?," *MedicineNet*, last updated September 10, 2015, http://www.medicinenet.com/probiotics/page3.htm.

12. Meghan B. Azad, Theodore Konya, Heather Maughan, David S. Guttman, Catherine J. Field, Radha S. Chari, Malcolm R. Sears, Allan B. Becker, James A. Scott, Anita L. Kozyrskyj, and the CHILD Study investigators, "Gut Microbiota of Healthy Canadian Infants: Profiles by Mode of Delivery and Infant Diet at 4 Months," *Canadian Medical Association Journal* 185, no. 5 (February 11, 2013): 385–94, doi:10.1503/cmaj.121189.

13. "Information for Pregnant Women," *Group B Strep Support*, accessed on September 7, 2016, http://gbss.org.uk/who-we-are/about-gbs/what-is-gbs/for-pregnant-women.

14. BabyCentre Medical Advisory Board (2012), "Group B Streptococcus in Pregnancy," *BabyCentre*, last updated on January, 2012, http://www.babycentre.co.uk/a1647/group-b-streptococcus-in-pregnancy.

15. Susan L. Fraser, "Enterococcal Infections," *Medscape*, March 17, 2016, http://emedicine.medscape.com/article/216993-overview.

16. Rebecca Dekker, "Group B Strep in Pregnancy: Evidence for Antibiotics and Alternatives," *Evidence Based Birth*, April 9, 2013, http://evidencebasedbirth.com/groupbstrep.

17. M. B. Azad, T. Konya, R. R. Persaud, D. S. Guttman, R. S. Chari, C. J. Field, M. R. Sears, P. J. Mandhane, S. E. Turvey, P. Subbarao, A. B. Becker, J. A. Scott, A. L. Kozyrskyj, and the CHILD Study investigators, "Impact of Maternal Intrapartum Antibiotics, Method of Birth and Breastfeeding on Gut Microbiota During the First Year of Life: A Prospective Cohort Study," *BJOG*, September 28, 2015 [Epub ahead of print], doi: 10.1111/1471-0528.13601.

18. David W. Hecht, "*Bacteroides* species," *Antimicrobe* database, accessed on September 7, 2016, http://www.antimicrobe.org /b85.asp.

19. Susan L. Fraser, "Enterococcal Infections," *Medscape*, last updated March 17, 2016, http://emedicine.medscape.com/article /216993-overview.

20. M. B. Azad, T. Konya, D. S. Guttman, C. J. Field, M. R. Sears, K. T. Hayglass, P. J. Mandhane, S. E. Turvey, P. Subbarao, A. B. Becker, J. A. Scott, A. L. Kozyrskyj, and the CHILD Study investigators, "Infant Gut Microbiota and Food Sensitization: Associations in the First Year of Life," *Clinical & Experimental Allergy* 45 (March 2015): 32–43, doi:10.1111 /cea.12487.

21. J. Smith, F. Plaat, and N. M. Fisk, "The Natural Caesarean: A Woman-Centred Technique," *BJOG* 115, no. 8 (July 2008): 1037–42, doi:10.1111/j.1471-0528.2008.01777.x.

22. "Impact of Maternal Intrapartum Antibiotics, Method of Birth and Breastfeeding on Gut Microbiota During the First Year of Life: A Prospective Cohort Study," *BJOG*, September 28, 2015.

23. Ewen Callaway, "Scientists Swab C-Section Babies with Mothers' Microbes," *Nature* [news], February 1, 2016, http:// www.nature.com/news/scientists-swab-c-section-babies -with-mothers-microbes-1.19275.

Chapter 5: What Is the Role of Bacteria in Training the Infant Immune System?

1. Patrice Nancy, Elisa Tagliani, Chin-Siean Tay, Patrik Asp, David E. Levy, and Adrian Erlebacher, "Chemokine Gene Silencing in Decidual Stromal Cells Limits T Cell Access to the Maternal-Fetal Interface," *Science* 336, no. 6086, June 8, 2012, 1317–21, doi:10.1126/science.1220030.

2. Ed Yong, "Newborn Immune Systems Suppressed," *The Scientist*, November 6, 2013, http://www.the-scientist.com /?articles.view/articleNo/38187/title/Newborn-Immune -Systems-Suppressed.

Chapter 6: How Is the Mother's Microbiome Passed on to Future Generations?

1. We made a documentary film called *Freedom for Birth* in 2012 (freedomforbirth.com) looking at the implications of a ruling by the European Court of Human Rights that every mother has the human right to decide the circumstances of her birth. Ternovszky v. Hungary, application no. 67545/09 (ECHR, December 14, 2010).

2. "What Did the Human Genome Project find?," Science Museum, http://www.sciencemuseum.org.uk/whoami/findoutmore /yourgenes/whatwasthehumangenomeproject/whatdidthe humangenomeprojectfind.

3. Robert Feil and Mario F. Fraga, "Epigenetics and the Environment: Emerging Patterns and Implications," *Nature Reviews Genetics* 13 (February 2012): 97–109, doi:10.1038/nrg 3142.

4. Helen Thomson, "'Epigenetic' Gene Tweaks Seem to Trigger Cancer," *New Scientist Daily News*, July 25, 2014, http://www .newscientist.com/article/dn25959-epigenetic-gene-tweaks -seem-to-trigger-cancer.

5. Robin McKie, "Why Do Identical Twins End Up Having Such Different Lives?," *The Observer*, June 2, 2013, http://www .theguardian.com/science/2013/jun/02/twins-identical-genes -different-health-study.

6. David M. J. Duhl, Harry Vrieling, Kimberly A. Miller, George L. Wolff, and Gregory S. Barsh, "Neomorphic *Agouti* Mutations in Obese Yellow Mice," *Nature Genetics* 8, no. 1 (1994): 59–65, doi:10.1038/ng0994-59.

7. Jill U. Adams, "Obesity, Epigenetics, and Gene Regulation," *Nature Education* 1, no. 1 (2008): 128, http://www.nature .com/scitable/topicpage/obesity-epigenetics-and-gene -regulation-927.

8. The EPIIC international research group was founded by Professors Soo Downe from the University of Central Lancashire, Holly Powell-Kennedy from Yale University, and Hannah Dahlen from Western Sydney University.

H. G. Dahlen, H. P. Kennedy, C. M. Anderson, A. F. Bell, A. Clark, M. Fourer, et al., "The EPIIC Hypothesis: Intrapartum Effects on the Neonatal Epigenome and Consequent Health Outcomes," *Medical Hypotheses* 80, no. 5 (May 2013): 656–62, doi: 10.1016/j.mehy.2013.01.017.

9. Malin Almgren, Titus Schlinzig, David Gomez-Cabrero, Agneta Gunnar, Mihael Sundin, Stefan Johansson, et al., "Cesarean Delivery and Hematopoietic Stem Cell Epigenetics in the Newborn Infant: Implications for Future Health?," *American Journal of Obstetrics and Gynecology* 211, no. 5 (July 1, 2014): 502e1–502.e8, doi:10.1016/j.ajog.2014.05.014.

Chapter 7: Is There a Link Between Cesarean Sections and Disease?

1. Noah H. Hillman, Suhas G. Kallapur, and Alan H. Jobe, "Physiology of Transition from Intrauterine to Extrauterine Life," *Clinics in Perinatology* 39, no. 4 (December 2012): 769–83, doi:10.1016/j.clp.2012.09.009.

2. Recent papers and media articles about the link between Cesarean sections and increased risk of certain health conditions:

    Jan Blustein, Jianmeng Liu, "Time to Consider the Risks of Caesarean Delivery for Long Term Child Health," *BMJ* 350 (June 10, 2015), doi:10.1136/bmj.h2410. For a lay summary see: Amy Kraft, "C-Section Births Linked to Long-Term Child Health Problems," *CBS News*, June 11, 2015, http://www .cbsnews.com/news/c-section-cesarean-births-child-health -problems-asthma-obesity-diabetes.

    Mairead Black, Siladitya Bhattacharya, Sam Philip, Jane E. Norman, and David J. McLernon, "Planned Cesarean Delivery at Term and Adverse Outcomes in Childhood Health," *Journal of the American Medical Association* 314, no. 21 (December 1, 2015): 2271–79. For a lay summary see: Roni Rabin, "C-Sections Are Best with a Little Labor, a Study Says," *Well* [blog], *New York Times*, December 14, 2016, http://well.blogs .nytimes.com/2015/12/14/c-sections-are-best-with-a-little -labor-a-study-says.

Astrid Sevelsted, Jakob Stokholm, Klaus Bønnelykke, and Hans Bisgaard, "Cesarean Section and Chronic Immune Disorders," *Pediatrics* 135, no. 1 (November 2014): e92–8, doi:10.1542/peds.2014-0596. For a lay summary see: Anne Ringgaard, "Giant Study Links C-Sections with Chronic Disorders," *ScienceNordic*, December 9, 2014, http://sciencenordic.com/giant-study-links-c-sections-chronic-disorders.

3. L. Huang, Q. Chen, Y. Zhao, W. Wang, F. Fang, and Y. Bao, "Is Elective Cesarean Section Associated with a Higher Risk of Asthma? A Meta-analysis," *Journal of Asthma* 52, no. 1 (August 27, 2014): 16–25, doi: 10.3109/02770903.2014.952435.

4. In research published in Germany, 2010, researchers found 28 percent of celiac children were born by Cesarean section compared with a maximum Cesarean section rate of 19 percent in all the other groups. Evalotte Decker, Guido Engelmann, Annette Findeisen, et al., "Cesarean Delivery Is Associated with Celiac Disease but Not Inflammatory Bowel Disease in Children," *Pediatrics* 125, no. 6 (June 2010): e1433–40, doi: 10.1542/peds.2009-2260. For a lay summary see: Jennifer Goodwin, "C-Sections May Raise Celiac Disease Risk in Offspring," *US News*, May 18, 2010, http://health.usnews.com/health-news/family-health/womens-health/articles/2010/05/18/c-sections-may-raise-celiac-disease-risk-in-offspring. See also: Katherine MacDonald, "Controversial Health Topic: Gluten Issues Are Caused by C-Sections," *Reader's Digest*, last accessed September 30, 2017, http://www.rd.com/health/wellness/controversial-health-topic-gluten-issues-are-caused-by-c-sections.

5. Blustein, "Time to Consider the Risks of Caesarean Delivery for Long Term Child Health."

6. "Asthma and Other Conditions," *Asthma U.K.*, http://www.asthma.org.uk/advice/manage-your-asthma/other-conditions. Alfredo A. Santillan, Carlos A. Camargo Jr., Graham A. Colditz, "A Meta-Analysis of Asthma and Risk of Lung Cancer (United States)," *Cancer Causes and Control* 14, no. 4 (May 2003): 327–34, http://www.ncbi.nlm.nih.gov/pubmed/12846363, PubMed PMID: 12846363.

Chapter 8: What Is the Impact on Humanity as a Whole?

1. "CDC Global Noncommunicable Diseases (NCDs)," Centers for Disease Control and Prevention, last updated May 24, 2016, http://www.cdc.gov/globalhealth/healthprotection/ncd.
2. "Noncommunicable Diseases" (fact sheet), World Health Organization, last updated January 2015, http://www.who.int/mediacentre/factsheets/fs355/en.
3. "2011 High Level Meeting on Prevention and Control of Non-communicable Diseases," United Nations, meeting on September 19–20, 2011, http://www.un.org/en/ga/ncd meeting2011.
4. Ibid.
5. *The Global Economic Burden of Non-communicable Diseases*, World Economic Forum (September 18, 2011), http://www.weforum.org/reports/global-economic-burden-non-communicable-diseases.
6. Ibid.
7. Kate Kelland, "Chronic Disease to Cost $47 Trillion by 2030: WEF," *Reuters*, http://www.reuters.com/article/us-disease-chronic-costs-idUSTRE78H2IY20110918.
8. "World Health Assembly Adopts New International Health Regulations" (press release), World Health Organization, May 23, 2005, http://www.who.int/mediacentre/news/releases/2005/pr_wha03/en.
9. Sara E. Davies, "National Security and Pandemics," *UN Chronicle* L, no. 2 (August 2013), http://unchronicle.un.org/article/national-security-and-pandemics.
10. Philip Shabecoff, "Global Warming Has Begun, Expert Tells Senate," *New York Times*, June 24, 1988, http://www.nytimes.com/1988/06/24/us/global-warming-has-begun-expert-tells-senate.html.
11. Suzanne Goldenberg, "US Senate Refuses to Accept Humanity's Role in Global Climate Change, Again," *The Guardian*, January 22, 2015, http://www.theguardian.com/environment/2015/jan/22/us-senate-man-climate-change-global-warming-hoax.

Index

181

Barker, David, 48
Barrett's Esophagus, 27
Bell, Aleeca
  empowering mothers, 152–53
  long term effects on humanity,
    142
  more research needed, 146
  team of experts, 9, 11–12
bifidobacteria
  functions, 52, 62
  importance for baby, 47
  mother's fecal matter, 42
  vaginal microbiome, 84, 104, 105
birth. *See* Cesarean section;
  premature/preterm birth; vaginal
  birth; vaginal microbiome
birth canal, 46, 61, 112
bisphenol A, 117
Blaser, Martin
  antibiotic winter, 141, 144, 152
  antibiotics, 30–31
  Cesarean section, 75
  child development, 53–54
  childbirth microbe seeding, 2, 45
  human microbiome, 21–22, 25,
    28–29
  increased risk of infectious
    disease, 140
  keystone species, 34–35, 112
  lactobacilli, 45–46, 61
  microbial diversity, 36, 148
  Missing Microbes, 2, 4–5, 11, 21,
    26, 28–29, 51, 61, 140, 146
  more research needed, 148
  pandemics, 140–41
  public awareness needed, 151
  rates of illness, 26–29, 36, 127
  reduced gut bacterial diversity,
    4–5

  researching causal studies, 129
  team of experts, 10–11, 13–14
  vernix, 51
  YouTube videos, 9
blood pressure, 138
blood-sugar levels, 138
Blustein, Jan, 127
BMJ, 127
Bolivia, 93
bonding, 57–58, 91
bowel disorders, 135
brain, gut microbiome relationship,
  24, 132–33, 136
Brazil, 30, 76
breast milk
  adapting to baby's changing
    needs, 63–66
  antibodies, 106
  bacteroides, 105
  bifidobacterium, 105
  colostrum, 60–61, 63, 68–69
  contents of, 60–61, 69
  gut microbes, 60, 147
  immune system development,
    105
  lactobacilli, 41, 105
  nutrients, 60, 68
  seed-and-feed process, 2–3,
    61–62
breastfeeding
  antibiotics, 91–92
  baby learning, 59
  bacteria differences in baby, 87
  Cesarean section, 67–68, 91–92
  difficulties with, 62–63
  gut microbes, 60, 147, 155
  lactobacilli, 41

Cabrera-Rubio, Raul, 63

About the Authors

Alto Films Ltd.

Toni Harman and Alex Wakeford are professional filmmakers who met at the London Film School more than twenty years ago. Since then they've been making films together.

Over recent years they have made four feature-length films that have been distributed internationally, including *Credo* (2008, released as *The Devil's Curse* by Lionsgate in the United States), a psychological thriller; *Doula!* (2010); and *Freedom for Birth* (2012), a documentary about human rights in childbirth.

Their most recent film, *Microbirth* (2014)—about how birth impacts a baby's lifelong health—won the Grand Prix Award at the Life Sciences Film Festival in Prague.

FSC
www.fsc.org
MIX
Paper from
responsible sources
FSC® C013483

green
press
INITIATIVE

Chelsea Green Publishing is committed to preserving ancient forests and natural resources. We elected to print this title on 100-percent postconsumer recycled paper, processed chlorine-free. As a result, for this printing, we have saved:

38 Trees (40' tall and 6-8" diameter)
17 Million BTUs of Total Energy
3,252 Pounds of Greenhouse Gases
17,638 Gallons of Wastewater
1,181 Pounds of Solid Waste

Chelsea Green Publishing made this paper choice because we and our printer, Thomson-Shore, Inc., are members of the Green Press Initiative, a nonprofit program dedicated to supporting authors, publishers, and suppliers in their efforts to reduce their use of fiber obtained from endangered forests. For more information, visit: www.greenpressinitiative.org.

Environmental impact estimates were made using the Environmental Defense Paper Calculator.
For more information visit: www.papercalculator.org.

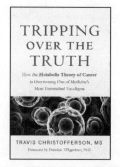
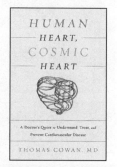

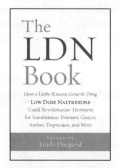
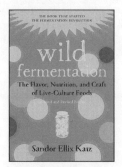
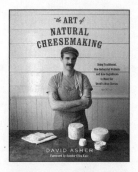